THE
AMERICAN
HOUSE OF
SAUD

THE AMERICAN HOUSE OF SAUD

THE SECRET PETRODOLLAR CONNECTION

STEVEN EMERSON

FRANKLIN WATTS 1985
NEW YORK TORONTO LONDON SYDNEY

The author gratefully acknowledges the permission
of the *New Republic* magazine to reprint portions
of Chapters Five, Eleven, and Fourteen.

Library of Congress in Publication Data

Emerson, Steven.
The American house of Saud.

Bibliography: p.
Includes index.
1. United States—Foreign relations—Saudi Arabia.
2. Saudi Arabia—Foreign relations—United States.
3. Lobbying—United States—History—20th century.
4. Corruption (in politics)—United States—History
—20th century. 5. United States—Foreign relations
—Near East. 6. Near East—Foreign relations—
United States. I. Title.
E183.8.S25E44 1985   327.73053'8   84-27067
ISBN 0-531-09778-1

Printed in the United States of America
6 5 4 3 2 1

CONTENTS

IN MEMORY OF
MY FATHER

ACKNOWLEDGMENTS

I am gratefully indebted to many people who have given me their time, ideas, and trust to make this book possible. Foremost are the numerous individuals in government, the business community, and elsewhere who risked jeopardizing the security of their careers by speaking to me; their courage and forthrightness will be always remembered and appreciated.

The origins of this book are a series of articles I wrote for *The New Republic* in 1982. I would like to express my appreciation to Charles Krauthammer, Dorothy Wickenden, Rick Hertzberg, Jim Glassman, Morton Kondracke, Martin Peretz, and others at the magazine for their critical guidance and suggestions in the preparation of my original articles and in subsequent editorial advice.

I have been the recipient of invaluable advice, suggestions, and feedback from close friends and colleagues. I am especially thankful to Jeffrey Jerred, Steven Schneider, Jerome Levinson, Elliott Negin, Richard Chimberg, David Welna, Chris Welna,

Neil Roland, Edwin Rothschild and Scott Armstrong. Jeffrey Robbins gave unselfishly of his time and energy from the beginning to the end; his critical scrutiny of the final manuscript proved to be of inestimable value. Jim Hertling was my research assistant during the summer of 1983. He displayed resourcefulness, investigative skills, and writing ability truly unique for someone his age. David Emerson's critical reading was invaluable.

I owe a special measure of gratitude to the law firm of Arnold and Porter and attorneys Irvin Nathan and Edward Wolf for representing me *pro bono* in my protracted legal battles with the Executive Branch over my requests for documents submitted under the Freedom of Information Act. Ed's determination and perseverance succeeded in prying loose hundreds of critical documents from the Department of State, Department of Justice, Department of Defense, Department of the Treasury, and the CIA.

I am grateful to the Project on Investigative Reporting on Money in Politics for providing me with a grant to undertake part of this investigation.

My agent, Julian Bach, stood behind me every inch of the way, and for that I will always be appreciative. I was very lucky to have had as my editor Ellen Joseph, whose qualities are those every writer should be treated to: always reassuring, a constant source of ideas, and gifted with an ability to penetrate to the core of the problem. Donna Ryan did extremely talented work in copyediting my manuscript.

Joan Shaffer, Sandy Echeverry and Joan Jordan are to be commended for typing various sections of the manuscript, toiling all hours of the day and night. A medal of honor must go to Sue Gruskiewicz for deciphering thousands of pages of my handwriting as she produced a final typewritten draft of the book. Sue's exceptional conscientiousness was matched only by her good humor and patience.

Finally, I would like to express my good fortune in having been able to work closely with the late Senator Frank Church, who died in early 1984. He was both a teacher and a friend, and it was the courage he displayed in his senate career as he fought both against the corporate power elite and governmental abuse of power and the bravery he displayed as he battled a vicious cancer that provided me with much of the inspiration to write this book.

August 1984

THE AMERICAN HOUSE OF SAUD

INTRODUCTION

The American House of Saud reveals how billions and billions of dollars in Arab oil money have secretly affected the political process in the United States. It discloses how Saudi Arabia and its influential corporate supporters, cultivated through the allure of money, have changed American foreign policy, manipulated the American public, and generated an impact in numerous other ways on American society.

That Saudi Arabia has sought to make its influence felt in the United States is not unique. Virtually every foreign government attempts to exert influence in the American political process. In Washington, each one of the foreign embassies serves as a de facto lobbyist in support of its government's interests.

Beyond the role of embassies, many foreign governments employ professional lobbyists to represent them at the appropriate level of the American government. This, too, is not unique. At the Department of Justice, over ten thousand individuals or groups are registered as agents of foreign governments as politically diverse as Canada, Nicaragua, South Africa, France, Argentina, Britain, Australia, and one hundred other countries. Registration is required of any American lobbyist representing or promoting the political interests of a foreign principal.

Japan, for example, operates one of the largest and best connected foreign lobbies in the United States, spending more money on its agents and activities than any other government. Of the more than one hundred foreign agents registered for Japan at the Department of Justice, scores are former U.S. government officials such as William Colby, former director of the CIA; Brock Adams, secretary of transportation in the Carter administration; and Richard Allen, former national security adviser in the Reagan adminstration. In addition to hiring these agents, Japan has also endowed special foundations and institutes designed to engender support for, or at least understanding of, its political objectives to ensure that American trade policy does not slam the door on the flow of Japanese exports to the United States (which outnumber American exports to Japan by a ratio of two to one). Other nations, like Japan, focus on narrow, though important, issues such as American tariff levels, Export-Import Bank credits, tourism, and income tax treaties.

Though Saudi Arabia and other Arab governments have specific bilateral policies which they promote, Arab governments also collectively try to achieve a more comprehensive and unparalleled objective: a certain degree of control over American Middle East policy. Arab governments have adopted a political agenda in the United States that encompasses a broad range of goals—from increasing American military sales to the Arab world to reducing American support for Israel.

The popular perception, however, is that the Arab lobby pales in significance when compared to the strength and effectiveness of its counterpart, the Israeli lobby—or the Jewish lobby, as it is also called. Indeed, mention the Arab lobby to a Washington politico and he or she will likely scoff at the notion that such a lobby exists; mention the Jewish lobby, however, and the politico will likely tell you how powerful it is.

I remember a brief discussion with a senior *New York Times* editor two years ago when I first started writing this book. I told the editor I was working on a book about lobbying on the Middle East. The editor responded, "It's about time someone wrote a book about the Jewish lobby."

The editor's assumption that Jews constitute the only Middle Eastern lobby is a belief widely shared. And with good reason. American Jews have in fact lobbied very visibly and successfully—for example, in marshaling congressional support for billions of dollars' worth of annual aid to Israel. The efforts of the Jewish community and its high-powered tactics are routinely reported in the media. One edition of CBS's "Face the Nation" in 1984 was devoted entirely to the question of whether Jews exercise too much influence in the United States. This was the only time in the previous two years that "Face the Nation" has ever focused exclusively on the actions of any lobby, domestic or foreign.

The American Jewish community has concentrated its political efforts—such as lobbying, congressional voting, and campaign contributions—on Congress because Congress is institutionally responsive to domestic constituencies.

Moreover, some American Jewish leaders have openly declared support for Israel their exclusive political agenda, and some candidates have done an excessive amount of pandering to the Jewish community. Extreme behavior of this kind reinforces the twin assumptions that (1) American support for Israel flows exclusively from the political muscle of the Jewish community and (2) American policy is blindly pro-Israel and anti-Arab. Who could not have reached these conclusions following the 1984 presidential primary season in which Democratic candidates Walter Mondale and Gary Hart tried to outbid each other in proclaiming their support for Israel, particularly for moving the American embassy from Tel Aviv to Jerusalem, Israel's capital.

In fact, both assumptions are inaccurate. Although American Jews are a powerful political force, Congressional support for Israel stems also from the fact that the American public overwhelmingly identifies with Israel—in part because of its democratic values and institutions and in part because of its reliability as a stable ally in a strategic area of the world. Even after Israel's siege of Beirut, televised nightly, in the summer of 1982, public opinion polls showed that Americans felt more sympathy for Israel than for the Arab nations. The American public also feels instinctively estranged from authoritarian, repressive, or religiously zealous regimes, the type of which has often characterized much of the Arab world.

Furthermore, contrary to common belief, American policy has often followed the dictates of Arab nations; yet Arab governments have been shielded from the accusation that they have intervened in American politics in large part because of the actions of the executive branch of the American government. In recognition of the lack of popular support for the Arab cause, successive presidential administrations since 1948 have agreed to initiate pro-Arab policies. In effect, the executive branch has adopted a dual foreign policy: it ceded to Congress the realm of U.S.-Israeli relations, but it took upon itself the responsibility of protecting the interests of Arab nations such as Saudi Arabia. The efforts of American Jewish congressional lobbies were forced into the open, whereas the Arab nations could sit back as the executive branch did their political homework.

In the early 1950s, for example, the Treasury Department conferred on American oil companies in Saudi Arabia the special "oil tax credit." That unique credit effectively allowed Saudi Arabia, rather than the U.S. government, to be the recipient of the oil companies' income tax payments. This arrangement obviated the President from having to ask Congress to authorize foreign aid for the Saudis.

More than thirty years later, the executive branch's dual foreign policy is still in force. Successive administrations have sold tens of billions of dollars of sophisticated weapons to Saudi Arabia without regard to that country's absorptive military capacity or whether those weapons are appropriate to Saudi Arabia's legitimate defense requirements. The most recent case occurred in early 1984. The Reagan administration proposed selling to Saudi Arabia 1200 shoulder-fired portable Stinger anti-aircraft missiles—despite the fact that the weapons are only marginally effective in protecting Saudi oil tankers. Moreover, Reagan officials, in briefing Congress, could not explain how they arrived at such a large number of these missiles to be sold to the Saudis, especially in light of the fact that such a sale would have diverted production from the American military. When Congress balked, citing concern for Israel's security and the possibility that the highly-mobile Stingers would fall into the hands of terrorists, President Reagan withdrew the proposal. A month later, the President unilaterally authorized the sale of the Stingers to Saudi Arabia under a rarely used "national security" legislative clause. What the public did not see in this episode—or in the Reagan administration's 1981 decision to allow the sale of AWACs to Saudi Arabia (see Chapter 10) or in many other incidents revealed in this book—was the long but invisible arm of the Saudi government.

The American public has also failed to realize that a substantial part of the Saudis' success in promoting their own interests has been achieved through the cultivation of a broad coalition of influential and powerful American supporters. In exchange for access to both the huge Saudi market—now the sixth largest U.S. export market—and Saudi petrodollar investments, American corporations, business executives, and former government officials have become ardent lobbyists for Saudi political interests. Across the United States, the diffusion of Saudi and other Arab petrodollar wealth has created a huge "petrocorporate" class.

The emergence of the pro-Saudi petrodollar lobby is a cause for concern, since it is a vested lobby with commercial interests tied to foreign governments, purporting to represent the American national interest. Very few members of the burgeoning petrocorporate class are registered as foreign agents; yet their interests are inextricably linked to the interests of Saudi Arabia and other oil-producing Arab nations. The Americans in the petrodollar lobby are motivated by the prospect of financial gain or by the fear of losing business with the Arab world. But the lobbyists cloak their political program in the American flag. What's good for Saudi Arabia has become good for Greyhound, Ford, Chase Manhattan, Westinghouse, Bechtel, and thousands of other corporations. In turn, what's good for these corporations has become good for the United States.

The American Jewish community's support of Israel, by contrast, is motivated by ideological and religious beliefs. Though I strongly disagree with many policies of the Israeli government and with the virtually uncritical endorsement that some American Jews extend to Israel, no one can challenge the integrity that induces their actions.

Similarly, no one can dispute the fact that the activities of the American Arab organizations in the United States are also motivated by sincere political convictions. For that reason, I will not focus on the role played by these organizations—except where it is linked to the petrocorporate class—in eliciting support for the Arab cause. Whether we agree or disagree with the political program of either American Jews or American Arabs is immaterial; both communities are using the system in perfectly legitimate ways to foster political objectives they believe to be in the best interests of the United States. And if one side succeeds over the other, that's the outcome of the American political process at work.

When a lobby that is controlled, directed, or influenced by a foreign nation attempts to secretly manipulate American policy, however, the integrity of the American decision-making process is subverted. In recognition of this precise problem a law was passed requiring all agents of foreign powers to register publicly as foreign agents and submit regular disclosure reports. No comparable registration or reporting requirements exist for domestic lobbyists.

The purpose of this book is to provide the American public with the facts about the Saudi Arabian influence in this country and with the dimensions of the secret petrodollar connection.

I conducted more than 500 interviews in doing the research for this book. Many of my sources occupy sensitive positions in government or the corporate business world. Whenever possible, I have tried to quote my sources by name. I also relied extensively on government documents obtained through the Freedom of Information Act and on internal government, corporate, and university documents provided to me by individuals connected with these institutions.

1

THE
SETTING

Friday, March 4, 1983, 10:30
A.M. ''The Congress is literally terrified of the most powerful
lobby in the United States which is called the Jewish community.'' All eyes are transfixed on the speaker as his resonant voice
booms throughout the reception hall. ''Congressmen will never
have the guts to take the position in opposition to the Jewish
community of America. . . . 'They [the Jews] will kill you with
the death of a thousand cuts.' ''

From his lectern, the speaker paints a terrifying portrait of
the power of American Jews, revealing their ''secret lobbying,''
their ''prominence'' in television, radio, and banking circles, their
manipulative ''control'' of American foreign policy, and their
vindictive ''sanctions'' against anyone who dares to disagree with
them.

At the conclusion of the forty-five-minute address, the audience bursts into applause, with a few people even cheering.

Had this event occurred in Damascus, Baghdad, or Tripoli, it might not have been considered unusual. But the event
took place many thousands of miles away from the Middle East.
The location was Birmingham, Alabama. And the speaker was

not some crazed extremist, but a widely known former congressman who served fifteen years in the House of Representatives, and who once ran for President, and who led the anti–Vietnam War movement.

On that particular Friday morning, former California Congressman Paul M. McCloskey revealed his special insights about Jews, American foreign policy, and the righteousness of the Arab cause to a discreet gathering of 150 businessmen, industrialists, and bankers. He beseeched them to become active lobbyists against Israel, predicting that such political action would be required if they wanted to trade with the Arab world. McCloskey was not alone in promulgating these views. Throughout the entire day, speaker after speaker—several of them ex–United States ambassadors—rose to condemn American support for Israel or excoriate the influence of American Jews in the kind of vitriolic language surely not heard in the South, or in any other part of the country, for a very long time.

Even more unusual was that this extraordinary conference was sponsored, financed, and hosted by *Fortune* 500 multinational corporations, prominent Alabama banks, the University of Alabama, and even a federally funded Alabama state agency. In attendance besides the businessmen from seven southern states were Arab diplomats and financiers, the chancellor and president of the University of Alabama, and investment bankers from Shearson/American Express, and E. F. Hutton. The mayor of Birmingham was also on hand to greet the participants.

What happened in Birmingham? Why would the University of Alabama, together with Birmingham banks and corporations, agree to sponsor a gathering marked by anti-Semitic invective? Why would a former ambassador defend Libyan dictator Muammar el-Qaddafi? Why would a well-respected former congressman, known for his commitment to social justice and his fight against bigotry, deliver before his southern audience a speech lambasting the "influence" and "control" exerted by the Jews? (Two weeks after McCloskey delivered his tirade against the Jews, he gave the keynote address at a congressional symposium in which he cited the need to fight against anti-Semitism). And why would other speakers describe American support for Israel as the exclusive result of the manipulative power of American Jews?

The answers to these questions revolve around one of the most intriguing political developments of our time: the story of how billions and billions of Arab oil profits—known as petro-dollars—have permeated the American political process. What happened in Birmingham is but one manifestation of the vast ripple effect of ugly influence peddling whose origins lie in the simple search for money and profits.

In Birmingham, local corporations seeking to solidify and expand their own business contacts with Saudi Arabia joined forces with a university in almost desperate search of Saudi funding. Together they formed the basic ingredients of the explosive conference. The conference—which was only one of a series of flagrant episodes over a ten-year period—may have been unique to Birmingham, but not to the rest of the country.

Across the United States, Arab oil money has set in motion a massive and sometimes uncontrolled domino reaction of surreptitious political manipulation, bigotry, and above all else, advancement of other nations' secret political maneuvering.

The roots of the emergence of Arab oil influence go back to the 1973 Saudi-imposed oil embargo. Though the embargo may have long ago faded in memory, its legacy is very much present today, for the coalescing of Arab oil producers in October 1973 resulted in a series of decisions whose repercussions have far outlasted the effects of the oil cutoff. The price of oil, which stood at $3 per 33-gallon barrel in October 1973, shot up overnight by 400 percent. Since that time, the price has risen to as high as $34 a barrel.

Over one and a half trillion dollars subsequently flowed into the treasuries of just 13 countries from the economies of over 130 nations. Never in modern history have so few nations been so enriched with such astounding wealth in so short a period of time. In 1970, for example, Saudi Arabia earned $2.3 billion in revenues from the sale of its oil. A decade later, it had collected more than fifty times that amount: its 1980 revenues exceeded $110 billion.

The collective total of Saudi earnings since 1973 is even more revealing. Between 1973 and 1984 Saudi Arabia earned $661 billion; but in the eighteen years prior to 1973, Saudi revenues amounted to less than $35 billion. (Of the $661 billion that flowed into the Saudi treasury in the last decade, over $50 billion came from the United States.) The other oil-producing nations enjoyed similar rags-to-riches experiences. From 1963 to 1973, the eleven largest members of OPEC earned $144 billion.[1] In the following decade the collective revenues earned by the same OPEC countries from the sale of oil amounted to $1,634,950,000.

The combination of oil leverage and fantastic financial wealth was immediately translated into political power. In fact, the new political realities were already in evidence before the October 1973 Arab-Israeli War had ended. America's European allies refused to allow refueling rights or even air space for the American resupply of military equipment to Israel. And, in the weeks and months following the October 1973 War, many black African countries severed relations with Israel directly as a result of pressure exerted by the Arab world.

By 1975, a new geopolitical balance was already emerging at the United Nations. In November of that year, the General Assembly overwhelmingly approved a resolution that condemned ''Zionism'' as a form of ''racism.'' Though many Israeli military actions have certainly deserved censure, the willingness of so many nations to declare the Jewish nationalist movement a form of evil was an indication of the Arab world's newly acquired persuasive powers. ''Zionism'' was transformed into a pejorative overnight.

In a short period of time, the Arab world had managed to make its obsession with Israel the singular concern of the United Nations and many of its agencies. In 1981, for example, almost half of the forty-five Security Council meetings were convened in order to act on Arab complaints against Israel. Not one of the forty-five Security Council meetings dealt with the Soviet invasion of Afghanistan, the Iran-Iraq War, or Libya's invasion of Chad. Respected nonpartisan U.N. organs such as UNESCO have become politicized as Arab pressure to ostracize Israel has been exerted.

The Europeans also fell victim to petrodollar pressure. In 1980, Britain, Italy, France, West Germany, and the other members of the European Economic Community—which would never contemplate negotiating with the Irish Republican Army, the Baader-Meinhof, or the Red Brigades—passed a resolution declaring that the Palestine Liberation Organization "must be involved in any Middle East peace settlement." Not one word was mentioned of the only peace treaty between Israel and an Arab country—the Camp David Accords.

Despite the fact that the PLO has provided training and weapons to neo-Nazis in Germany and European terrorist groups and openly taken credit for terrorist attacks on Israeli and Jewish targets, the PLO was authorized to open official offices in almost every European capital.

Even newly elected French President François Mitterrand, a socialist who had maintained long and deep ties with Israel, became acutely aware of how vulnerable his government could be to Arab financial leverage. One of his first official acts was to dispatch an emissary to various Arab governments to reassure them that French weapons sales would continue uninterrupted. Moreover, his longtime championing of Israel's right to be recognized as the basic premise of any peace settlement soon fell by the wayside. According to the *Washington Post,* in the spring of 1982, Saudi Arabia—which had a trade surplus of $10 billion with France the previous year—deliberately refused to support the French franc, "contributing to a politically embarrassing French currency devaluation and to Mitterrand's [awareness] of the need to mend Arab fences."[2] Shortly afterward, Mitterrand afforded the PLO new international standing and respect by describing it as a "resistance organization." That same year Douglas Hurd, a deputy to British Foreign Minister Secretary Francis Pimm, became the highest ranking British official to meet with the PLO. And in the area of arms sales, the European democracies have sold more than 30 billion dollars' worth of the latest military equipment to the Arab world without any recognition of the destabilizing effects of these arms, the limited absorptive capacity of the Persian Gulf countries to assimilate these weapons, and the likelihood—as evidenced by the indiscriminate shelling in the Persian Gulf in the Iran-Iraq war—that these weapons will be used one against the other.

In Japan, PLO leader Yasser Arafat was received in October 1981 with the honors and protocol usually afforded to heads of state. Despite the fact that the PLO had aided the Japanese Red Army, a terrorist group which killed twenty-six people in a machine-gun attack in Israel in 1972, the PLO leader was honored by Japanese Prime Minister Zenko Susuki and welcomed by leading Japanese corporations in full-page newspaper advertisements.

The United States has been conventionally thought of as immune to the pressure of petrodollar politics for a variety of reasons, including the existence of a large and influential Jewish community, the general sympathy for and identification with the democratic values nurtured by Israel, the continuously high large amounts of aid provided by Congress to Israel, and the pluralistic insulation afforded by this country's immense social and economic diversity.

The reality, however, is far different. The enormous amount of Arab investment in and trade with the United States has produced dramatic shifts in American policy, and has spawned the emergence of an entire new petroclass of American supporters of Arab interests. Spanning the entire range of the political spectrum and corporate community, prominent American institutions—both governmental and private—have become vehicles for the furtherance of the Arab world's political aims.

To be sure, Arab investments in the United States—now estimated to be between $100 and 150 billion—and Arab purchases of American goods and services—which totaled $14 billion in 1983—are not unusual. Indeed, Arab investments constitute less than 10 percent of all foreign investment in the United States; and trade with the Arab world represents less than 10 percent of all American exports. But a critical distinction has rendered petrodollar trade different from trade with other foreign nations: Unlike other foreign investments and export markets, such as those of Europe or Japan, trade with the Arab world often comes with implicit political strings. Decisions regarding investments and trade are made by Arab leaders, government agencies, royal families, and businessmen. And because these decisions are made by the ruling elite or by officials inextricably tied to the government, Arab financial transactions have often been accompanied by a political dimension tied to foreign policy interests. Thus, unlike European transactions, which are made by private citizens with the exclusive economic purpose of profit or gain, the recycling of Arab oil dollars in the United States has often provided a conduit for the furtherance of Arab political goals.

In the end, a vast ripple of petrodollar influence has washed over American society as business and politics have become intertwined. Not only has Saudi Arabia, as the largest and most powerful oil producer, been able to directly change American policy, but it has been able to cultivate supporters from all walks of American life. Tremendous leverage is wielded over companies, consultants, officials, and institutions doing business with Arab oil producers. Banks, law firms, energy and munitions companies, construction firms, retail manufacturers, farmers, think tanks, rice growers, bus companies, ambassadors, farm co-ops, public relations firms, and even former Presidents have all become involved in varying degrees in pressing for the political interests of the Arab oil countries. Even prominent newspapers, universities, and television networks have become witting and unwitting vehicles—through being adroitly manipulated—for the promotion of the Arab political agenda.

To the unsuspecting, the emergence of across-the-board "understanding" and uncritical support for Arab political interests—ranging from arms sales to television commentaries to university programs—and simultaneous (and selective) condemnation of Israel and the American Jewish community, seems, for the most part, to be spontaneous. This appears as a function of disinterested political analysis, genuine intellectual and cultural inquiry, and an honest, objective assessment of the moral grays that dot the political landscape of the Middle East.

But, in fact, a great deal of the sudden "understanding" for and political support for Arab policies has little to do with independent political evaluations or even with exasperation at Israel's aggressive actions. The root cause is money. And through surreptitious political funding, manipulation of the political process, invisible political strings, hidden lobbying, and the dissemination of disinformation, the hidden hand of a vested petrodollar lobby has successfully and secretly affected the conduct of the American political process.

Ironically, the willingness of American corporations and executives to do the bidding of Arab governments constitutes a reversal of the traditional—if unethical—role of multinational corporations. For years, American and other multinationals have broken international laws or have interfered in the affairs and policies of foreign governments. United Fruit helped bring about the 1954 coup d'etat in Guatemala. ITT helped destabilize the regime of Chilean President Salvador Allende in 1973. And Mobil Oil supplied oil to Rhodesia in the 1970s in violation of the sanctions imposed by the U.N. Security Council. Petrodollar trade, however, has created a reverse imperialism: Instead of trying to advance American "interests" abroad, American corporations here become the pawns of Arab nations in trying to change American policies.

In the eleven years since the Arab oil embargo, the oil weapon has been transformed into a money weapon. Arab oil dollars have become thoroughly diffused throughout American society, yet they have retained their political strings. From the Ford Motor Company to General Electric, from former Vice-President Spiro Agnew to former President Gerald Ford, from Duke University to the University of Southern California, from Alabama Public Television to *Time* magazine, from the North Dakota Legislature to the U.S. Senate, the influence of petro-dollars has successfully—and secretly—changed the course of American politics. This is that untold story.

2

THE
LOYALTY
TEST

The story of Aramco is the story of
by far the greatest, most important,
and most dramatic overseas American
enterprise which has ever existed in
the history of our country to this date.

G. William Parkhurst,
former vice-president,
Standard Oil, in
testimony before the
Senate Subcommittee on
Multinationals, 1974

After flying into Geneva from their respective corporate headquarters around the world—New York, California, and Saudi Arabia—seven oil company executives arrived on May 23, 1973, at the Geneva Intercontinental Hotel for a meeting with the Saudi oil minister, Sheikh Ahmed Zaki Yamani. The purpose of the meeting was to discuss the impending negotiations between Aramco and Saudi Arabia over the transfer of the company's ownership to its Saudi hosts. But before the meeting got under way, Yamani asked his visitors whether they would mind paying a courtesy visit to King Faisal, the seventy-year-old monarch, who had also been staying at the hotel and resting from a tiring visit to Paris and Cairo.

Expecting to exchange only a few pleasantries, the oil company officials were surprised when the king launched into a bitter attack on American foreign policy in the Middle East. A few minutes later, the king issued an extraordinary threat: the oil company executives "would lose everything"—their oil concession—unless they acted immediately to "inform" the American public of its "true interests" in the Middle East.

From the moment this meeting ended, a new chapter had begun in the relations between Aramco and Saudi Arabia. The relationship now assumed a political dimension that would far surpass all previous acts of political subordination by Aramco to its Saudi hosts. Indeed, the ease with which King Faisal elicited the cooperation of the American oil companies in his efforts to change American policy, coupled with the immense wealth his country began to amass later that year, set the standard for the future behavior of hundreds of American corporations operating in Saudi Arabia and other Arab oil-exporting countries.[1]

It had been no accident of fate that Aramco, of all the hundreds of multinational corporations existing worldwide, would be the one multinational to exhibit a subservience to Saudi political needs infrequently paralleled in the history of multinational corporations.

Faced with bankruptcy in 1933 because of the decreased number of Muslims making pilgrimages to Islam's holiest city of Mecca—Saudi taxation on Muslim pilgrims had provided its main source of revenue—King Ibn Saud was forced to seek other sources of revenue for the one-year-old kingdom of Saudi Arabia. So in 1933, the king granted exploration rights to Standard Oil of California (SoCal). Three years later, oil was struck at Damman No. 7, located in the eastern provinces of Saudi Arabia. The discovery of oil in that particular well eventually led to the formation of a giant multinational oil consortium, the Arabian American Oil Company, or Aramco. In 1936, Texaco joined SoCal, and by 1944, the name Aramco had been coined and adopted. Two years later, Mobil and Exxon were allocated parts of the Saudi concession and also joined the Aramco consortium.

By the early 1950s, Aramco had established itself as one of the most important corporations in the world by virtue of the fact that it produced 90 percent of Saudi oil—a position it continues to occupy to this day. By the end of 1982, the company employed over fifty thousand people. Its gross receipts were over $50 billion—a figure that then exceeded the budgets of over one hundred nations of the world.[2]

In 1948, oil company officials actively opposed the creation of the state of Israel and, together with the U.S. State Department and Defense Department, pressed the Truman administration not to recognize the Jewish state for fear of offending the Arab countries, particularly Saudi Arabia. Oil officials collaborated shortly again with the National Security Council and the Treasury and State Departments to secure an agreement that allowed the oil companies to treat their payments for oil to Saudi Arabia as income tax payments rather than royalties, which is what the payments really were. This enabled the companies to reduce their American tax liability dollar for dollar for the "taxes" paid to Saudi Arabia. This agreement, which took effect in 1951, had the practical effect of shifting American oil company tax payments from the U.S. Treasury to the coffers of Saudi Arabia, thus obviating the need to provide overt foreign aid to the kingdom.[3]

In 1952, when the Saudis occupied Oman's Buraimi Oasis, Aramco provided transportation for the Saudi troops.[4] Aramco also served as King Saud's culinary caretaker, preparing his meals and hiring his chefs in the 1940s and 1950s. In an extension of their policy of boycotting anything Jewish, the Saudis prohibited Jews from working for Aramco in Saudi Arabia. Aramco implemented this exclusionary policy in a convenient way. It required all job applicants to obtain a visa from the Saudi government. Religious declaration was a required entry on Saudi visas and, in turn, any Jewish applicant was routinely rejected. (In 1959, a celebrated court ruling in New York rejected the right of Aramco to inquire directly about the applicant's religion. In decrying the "film of oil which blurs the vision of Aramco," the court ruled that "(o)ur basic documents of freedom are never to be subordinated to immediate business gain . . . no matter what the King of Saudi Arabia says . . .")

Throughout the 1950s and 1960s relations between Saudi Arabia and Aramco were as harmonious as ever. But the beginning of the next decade brought profound turbulence. The troubles started in Libya but spread rapidly throughout the Persian

Gulf. In 1970, newly installed Libyan dictator Muammar el-Qaddafi, who had ousted pro-American King Idris, demanded a radical increase in the price of oil from the oil companies in his country, Exxon and Occidental. But the companies resisted, and Qaddafi decided to make a test case: He ordered Occidental, which had depended on its Libyan operations for all of its non-U.S. profit, to slash production by 50 percent. By the end of 1970, Libya had won its demands for higher oil company payments—and in a domino reaction, the other Persian Gulf producers demanded and received similar increases.

In the next two years, the oil companies came to terms on a series of agreements with the newly cohesive OPEC cartel, only to witness individual countries like Iran and Libya break the terms and instigate new rounds of price leapfrogging. The success of the OPEC producers in eliciting concessions from the companies gave way to a new self-confidence: in addition to price increases, the producing countries began to demand partial control of the assets of the oil companies. And in a terrifying demonstration of their determination and power, Iraq and Libya nationalized two companies operating on their soil, the Iraq Petroleum Company and British Petroleum.

Agreement was finally reached by October 1972 that called for an immediate transfer of 25 percent of each oil company's assets to the host oil producing country. This was to be followed by a graduated transfer of the remaining assets until 51 percent was obtained in 1983. Known as the General Agreement, it also set a price schedule for oil company purchase of OPEC oil; but it left open to the companies and host governments the exact details of the transfer of ownership and the specific future role to be played by the companies and producing countries.

In Saudi Arabia, the oil companies in the Aramco consortium were not alarmed by the impending transfer of assets; their only concern was maintaining their rights to market and sell Saudi crude—an issue that had never been in dispute with the Saudi hosts. In fact, the oil companies knew that Saudi ownership would prove more profitable for them because it meant that the expensive, planned increase in productive capital capacity would be paid not by the companies but by the Saudis.

When negotiations between Aramco and Saudi Arabia commenced in December 1972, however, Yamani immediately upped the ante on a tangential, but important, issue. He demanded from the oil companies more money per barrel of oil than had been agreed on in the 1972 General Agreement. It was clear that the Saudis were swept up in the fervor of pan-Arab nationalism that would reach a new crescendo every time OPEC forced the oil companies to capitulate to a new set of demands. Political leapfrogging became unavoidable.

Not surprisingly, an uneasy tension began to develop beneath the calm veneer of the Aramco-Saudi relationship. Disagreements emerged over price and also over the much more sensitive subject of oil production levels. Some Saudi officials charged that Aramco was deliberately pumping certain oil fields, especially those that contained the favored Arabian "light," beyond geologically prudent rates in an attempt to extract as much oil as possible prior to nationalization. At one point, the Saudis ordered Aramco to curtail production severely in their third largest oil field, Abqaiq, because of the structural damage to the reservoir that resulted from excessive pumping. Still, even as other producing countries like Iraq and Libya had nationalized some companies, the threat of expropriating Aramco's assets or jeopardizing its oil concession had never come up in the discussions between the Saudi leaders and the oil men.

While OPEC radicalism was rising, King Faisal, the austere Saudi leader who succeeded to the throne in November 1964, was becoming increasingly bellicose toward the United States and Israel. Faisal had never reconciled himself to the idea of a Jewish state and would routinely distribute to visitors copies of the notorious anti-Semitic forgery, the *Protocols of the Elders of Zion*. In April 1973 Faisal ordered Sheikh Yamani to Washington to warn the Nixon White House that unless Washington forced Israel to immediately quit the territory—including Jerusalem—captured in the June 1967 War, Saudi Arabia would not maintain its current oil production levels. As a result of the dramatic tightening of the world oil market, Saudi oil was becoming increasingly vital to the security of the West. In just three years, between 1970 and 1973, the Saudi share of the world oil market skyrocketed from 12.8 percent to 21.4 percent.

Increased demand for Saudi oil and the oil of other OPEC members emboldened the Arabs to link oil and politics together for the first time in a credible threat.[5] Kuwait Foreign Minister Abdul Rahman Atiqi warned the West in January 1973 that his country was "prepared to use oil as a weapon against Israel." On May 15, Libya sent shock waves around the world by suspending all oil exports for a full day, creating a symbolic precedent for its Arab neighbors and a powerful reminder to its Western customers.

It was no wonder then that American oil executives in Saudi Arabia became extremely nervous about the open talk of the oil weapon, especially since their relations with the Saudis had already been jarred by disagreements over price and production. Given the mercurial changes in the king's moods, who knew what he might do in a fit of rage?

With greater frequency the king began to lash out at Frank Jungers, the mild-mannered president of Aramco, on the subject of American policy and communism, which the king equated with Zionism. One meeting, which occurred in early May 1973, left a definite impression on Jungers. Protocol had required that Jungers, the second highest Aramco official, pay routine courtesy visits to the king. Arriving on May 3 for such a visit, Jungers discovered Oil Minister Yamani and Foreign Minister Ommar Saqqaf also in attendance. The meeting was not going to be business as usual. He was right. In blunt language, Faisal demanded that the oil companies change American policy. He repeatedly warned that it was up to the American companies, if they wanted to preserve their interests, to prove their loyalty to Saudi Arabia. The session lasted for thirty minutes. Immediately afterward, Jungers cabled officials of Texaco, Mobil, Exxon, and Standard Oil in the United States with a summary of the king's admonitions. A wave of anxiety swept through oil company officialdom, but there was still no cause to press the panic button.

That time came two and a half weeks later in Switzerland. In his Geneva hotel suite on May 23, King Faisal met with the Aramco directors. The king told them that although he had received "assurances of [political] cooperation from France," he insisted that "this was not enough." He complained bitterly of the "failure of the United States government to give Saudi Arabia positive support." Then, for the first time in Aramco's history, the king unambiguously warned them that Aramco's oil concession was in jeopardy. The only way to save it was to attempt to redirect American policy away from support for Israel. "You may lose everything," the king said, "time is running out." Six days later, one of the oil officials attending the meeting— W. J. McQuinn, a vice-president of Aramco—reported the substance of the king's warnings in a confidential memorandum distributed to the oil companies' U.S. offices on May 29:

> *You will lose everything (concession is clearly at risk). Things we [the oil companies] must do: (1) inform U.S. public of your true interests in the area (they are now being misled by controlled news media) and (2) inform government leaders—and promptly*

In the next few days, top executives of the four Aramco companies authorized the necessary political action. Their oil concession had to be saved. To be effective, they had to portray, in all future lobbying endeavors, the king's threats to their oil concession as a threat to American national interests. So, those officials who had attended the Geneva meeting flew to Washington to lobby U.S. government officials in a series of meetings on May 30. The contingent included A. C. DeCrane of Texaco, Charles J. Hedlund of Exxon, Henry C. Moses of Mobil, W. J. McQuinn of Standard Oil, and J. J. Johnston of Aramco.

According to an Aramco cable, the meetings lasted an entire day at the White House, State Department, and Pentagon. At 10:30 A.M., the contingent met with Assistant Secretary of State Joseph Sisco, one of Henry Kissinger's top deputies on the Middle East. At 2:30 P.M., the oil company executives met at the White House with General Brent Scrowcroft and other aides of the National Security Council and at 4:00 P.M., the group met at the Pentagon with Acting Defense Secretary Bill Clements. At each of the sessions, the oil company contingent relayed the warnings of King Faisal that if American policy did not change, then "all American interests in the Arab world will suffer."

Throughout the summer the oil companies ensured that American policymakers were exposed to a steady drumbeat of the same ominous warnings. Exxon even appointed an official to provide regular briefings to State Department staffers. By early August, those involved in the lobbying expanded beyond the traditional coterie of oil company executives with direct business interests in the kingdom of Saudi Arabia.

One of the lobbyists was John J. McCloy, a senior partner in the New York law firm of Milbank Tweed Hadley & McCloy, who was considered one of the most powerful lawyers in the United States. McCloy had served in numerous positions of power and influence, including U.S. high commissioner for Germany during World War II, chief executive officer of the Chase Manhattan Bank, chairman of the Ford Foundation and president of the World Bank. An adviser to Presidents Roosevelt, Truman, Eisenhower, and Kennedy, McCloy by 1962 had earned a place as the leader of the "American Establishment," as noted by *New Yorker* columnist Richard Rovere.

On August 9, 1973, McCloy accompanied senior executives of ten oil companies (Continental, Exxon, Gulf, Nelson Bunker Hunt, Texaco, Arco, Mobil, Marathon, Occidental, and SoCal) for a meeting with Under Secretary William Casey, five other State Department officials, and a representative from the Justice Department. The 4:30 P.M. meeting took place in a special conference room on the State Department's seventh floor, which also houses the office of the secretary of state. The ninety-minute session was initiated by Casey, who had requested that McCloy provide a briefing so that the State Department could "be brought up to date on developments in Libya and generally throughout the Middle East producing countries." Though much of the meeting was devoted to a discussion of Muammar el-Qaddafi's threats to nationalize American oil companies operating in Libya, McCloy also gave his thoughts on the Arab–Israeli conflict. He told the group that "progress in the Arab–Israeli dispute was becoming daily more important, particularly in light of Faisal's change of position." And he warned that "our standing with the Arab world was becoming weaker each day."

Meanwhile, in order to shore up their standing with key Saudis, Aramco officials began to provide special briefings to visiting American officials, investors, and reporters. At these briefings the oil executives emphasized the importance of solid American-Saudi friendship, the need to be responsive to Saudi political interests, and the need to recognize the dangers to "American national interests" that attended support for Israel. Jungers sent cables to the four parent companies revealing how he succeeded in weaving in subtly the more politicized discussion on the Arab–Israeli dispute: "We found it a good tactic to concentrate on the energy crisis and American interests in the Middle East and to treat the Arab–Israeli problem as peripheral to these." In those briefings, Israel was depicted as the sole source of instability in the Middle East and, as such, was held singularly responsible for the "deterioration of U.S. influence in the Arab World." Israel was even held solely accountable for the "burgeoning" of Soviet influence in the Middle East. The scores of visitors to Saudi Arabia who were treated to such briefings included General Andrew Goodpaster, the Supreme Allied Commander for Europe, and Rear Admiral Earl R. Crawford, deputy chief of staff, Allied Command.

One visitor who commanded special attention was John J. O'Connell, a vice-president of the multinational construction firm, Bechtel. Unlike other visitors, O'Connell was provided with a private briefing paper upon his return to the United States. Aramco officials duly noted in internal memoranda that O'Connell was planning to spend Nixon's inaugural day, January 20, 1973, with Secretary of the Treasury George P. Shultz, who would soon become head of the Bechtel Corporation, a huge multinational company with billions of dollars worth of construction projects in Saudi Arabia. Aramco's instructions called for a three-page paper to be hand delivered to O'Connell "with no attribution or Aramco identification." The oil company officials were apparently hoping that O'Connell would discuss the paper with Shultz or show it to him.

In this memorandum as well as in all other oil company lobbying efforts and communications, Aramco talked only in lofty foreign policy prose. It was a tactical imperative for the oil companies to claim they were protecting the "national interest." Who could argue with patriotism?

Aramco's efforts soon switched into high gear. On June 23, 1973 Mobil published a "commentary" on the Middle East in the *New York Times* in the traditional Mobil "Observer" style. Though Mobil had previously published quarter-page commentaries on various domestic issues, this was the first venture into foreign policy. Citing the "U.S. stake in the Middle East," Mobil warned against allowing American relations with the Arab world to deteriorate. (This was the first of a series of Middle East newspaper advertisements that Mobil published over the next ten years.) Besides direct Saudi pressure, Mobil had even a narrower self-interest in engendering Saudi goodwill. In the early years of the Aramco consortium, Mobil had been allocated—by deliberate request—the lowest percentage of Saudi oil production. With hindsight, the company realized its mistake and had been clamoring for a larger share.

In the week before Mobil placed its ad, Exxon initiated its own pro-Saudi efforts, albeit on a much lesser scale. Exxon vice-president Howard Page delivered a favorable speech to the Alumni Association of the American University of Beirut in New York City. On June 26, three days after the Mobil ad appeared, SoCal's

McQuinn cabled George Keller, vice-chairman of Aramco confirming the plan of action:

> As noted briefly in the Foreign Review Committee meeting June 25, we need to take some positive action whereby we can demonstrate to SAG (Saudi Arabian government) that we are not unmindful of their interests and problems. The recent advertisement of the New York Times by Mobil and the speech by Howard Page to the Alumni Association of the American University in Beirut are examples of what can be done. These actions by Mobil and Exxon will be brought to SAG's attention by Aramco.

But McQuinn also cautioned Keller that the absence of similar political action by Standard Oil would appear conspicuous to Saudi leaders. He suggested therefore that Otto N. Miller, chairman of Standard Oil, send a letter to his company's employees detailing the "need to conduct our foreign affairs in a manner that will assure reasonable access to foreign oil." McQuinn urged that Miller be directly apprised of his concern—and that Miller's endorsement be obtained to a speedy "course of action."

On July 10, the oil companies received their anxiously awaited answer from the Saudis. In a meeting held at Aramco's headquarters in Dhahran, Sheikh Yamani, the oil minister, told Frank Jungers, Aramco's president, that he had seen the Mobil ad and had been asked by the king to write a letter to Mobil commending the company for its "positive step." Yamani added, however, that Mobil's actions should be considered "just a beginning" of what the companies had to do in the future.

In the meantime, Standard Oil commenced its political offensive. On July 26, exactly one month after Aramco cited the need for Standard Oil's involvement, Chairman Otto N. Miller publicly urged the United States to reexamine its policy toward the Arab world in a letter addressed to the company's 260,000 stockholders and 40,000 employees. He urged his audience to help press for the U.S. government to "work more closely with the Arab governments to build up and enhance our relations with the Arab people." He concluded with the following plea: "There must be understanding on our part of the aspirations of the Arab people and more positive support for their efforts toward peace in the Middle East."

One week later, on Thursday, August 2, McQuinn handed a copy of Miller's stockholder letter to Yamani. Hastily glancing at it, Yamani complained that since it had gone only to stockholders, American press attention would be minimal. McQuinn assured him there would be widespread press publicity. Yet five days later, the press coverage had been sporadic. McQuinn gathered all the available press clippings for delivery to Yamani and also urged his colleagues in New York to advise Aramco to arrange through its contacts to have Miller's letter published in the London-based magazine, *Middle East Economic Survey.*

Suddenly, McQuinn's assurances to Yamani were realized. Across the United States, American Jews condemned Miller's letter. Now that the Jews had acknowledged they had been stung, Miller's action became news: his letter had extraordinary impact throughout the Arab world. On August 5 and 6, Saudi and other Arab newspapers gave front-page attention to the letter. The Beirut daily, *Al-Liwa* for example, carried the following headline on its front page: "Signs of Change in American Public Opinion in Favor of Arab Cause." A few days later, to the delight of Aramco's officials, Miller's action continued to draw press attention. On the evening of August 9, Otto Miller was scheduled to deliver a speech in New York City. Inside the hotel where he was about to speak, members of the radical Jewish Defense League heckled him, resulting in some members being arrested. Pictures of screaming young Jewish demonstrators were broadcast over the television networks. It was now a full-blown media event, especially since it had happened in the media capital of the world.

Aramco officials were ecstatic. In a confidential cable dated August 13 with instructions for "special handling" sent to J. J. Johnson, vice-president of Aramco in New York, Aramco vice-president R. W. Powers boasted of the "excellent press and radio coverage in the Middle East and Saudi Arabia" given to Miller's letter and the ensuing arrests of the demonstrators: "Complete coverage of the letter was carried by most Arab radios and the arrest of Jewish Defense League activists in New York through August 9 was played up by radios Riyadh and Jidda, as well as by other Arab radios." Powers also noted that "Miller's letter coming as it does so closely on the heels of the Mobil ad (which was also most effective) has been welcomed as a real step forward by Saudi and other Arabs we have talked to. Perhaps, more important, these efforts have had excellent impact on King Faisal, Yamani and others." The "reaction in Saudi Arabia," he added in the final sentence of the cable, has been "most encouraging and indicates that additional effort(s) along similar lines would be most useful."

Aramco's activities had the desired effect: King Faisal was mollified. Terminating the oil concession was no longer dangled as a sword of Damocles. But the king was still threatening to use oil as a weapon against the West and the United States. On August 23, President Anwar Sadat of Egypt flew to Riyadh for a secret meeting with Faisal. Though no documentation exists of what transpired at the meeting, several Middle East observers believe that Sadat informed Faisal of his intent to attack Israel and asked the Saudi monarch for financial support as well as his willingness to utilize the oil weapon. Whatever transpired between the two men, immediately following Sadat's departure Faisal warned the United States that Saudi Arabia would limit its oil supplies: "America's complete support of Zionism against the Arabs makes it extremely difficult for us to continue to supply the U.S. petroleum needs."

In the month of September, President Sadat abruptly moderated his anti-Israel rhetoric and announced that all reservists who had been mobilized would be released by October 8, 1973.

These actions stood in startling contrast to repeated mobilizations of the Egyptian armed forces during the previous six months, the civilian sector practice blackouts, and the anti-Israel rhetoric of the government. Then the signals to Israeli and American intelligence were reversed: the evidence was that Egypt was not preparing for war. In fact, in early October, Sadat indirectly informed Secretary of State Henry Kissinger that his army maneuvers were merely practice exercises and that Egypt was prepared to open negotiations with Israel.[6]

But on October 6, the Jews' holiest day of the year—Yom Kippur, the Day of Atonement—Egypt and Syria assaulted Israeli forces in one of the most strategically destructive offensive attacks in modern military history. For the first three days of the Yom Kippur War, it looked to Israeli leaders that their ghastly nightmare—being overrun by the Arabs—might become real. Syrian tanks overtook all Israeli positions on the Golan Heights and were poised to invade northern Israel. Prime Minister Golda Meir ordered that plans be readied for the evacuation of the Israeli population in the North. On the western front, Egyptian tanks blasted through the paper-thin Bar-Lev line and established a fourteen-mile beachhead. Israeli Defense Minister Moshe Dayan, according to various Israeli military historians, was ready to accept defeat. The Arabs were euphoric. Official Arab radio stations declared that the Jews would be finally "thrown into the sea."

Six days after the outbreak of the war, during which Israel was in a state of shock from the loss of two thousand soldiers, the American oil moguls leaped into the fray. On October 12, before any effort to resupply Israel had been initiated, the chairmen of Exxon, Mobil, Texaco, and Standard Oil felt compelled to press the Arab case before President Richard Nixon. A cover letter written by John J. McCloy to Nixon's aide, General Alexander M. Haig, stated that the "chief Aramco shareholders with large concessionary interests in Saudi Arabia wish to have this brief memorandum summarizing the critical situation in regard to the flow of oil supplies from the Middle East placed in the President's hands as soon as possible." In their "Memorandum to the President," the four chief executives warned against "any further demonstration of increased U.S. support for the Israeli position." They asserted further that "increased military aid to Israel will have a critical and adverse effect on our relations with moderate Arab oil producing countries."

The Nixon administration wavered in its initial deliberations over whether to supply Israel with the military equipment it needed to defend itself. Already, all of Western Europe, with the notable exception of the Netherlands, had frozen all shipments of military equipment previously ordered by Israel. Moreover, each country announced that the United States would not be granted base refueling or even air refueling rights in any effort to resupply Israel. Finally, however, on October 13, the United States announced that it had decided to resupply Israel. But deliveries did not actually begin until October 20.

On the battlefront, the Israelis began to launch counterattacks. Focusing initially on the Syrians, whose proximity to Israeli population centers posed the greatest threat, Israeli forces regrouped south of the towering Golan Heights. In the largest single tank battle since Field Marshal Erwin Rommel confronted the Allies in northern Africa during World War II, the Israelis recaptured the Heights and began their drive on Damascus. Stopping seventeen miles from the Syrian capital, the Israelis shifted their attention to the south. There they attempted but failed to dislodge the Egyptians from the beachhead they had established on the eastern bank of the Suez Canal. But on October 16, the maverick Israeli General Ariel Sharon broke through Egyptian lines on the other side of the canal and was soon on his way toward isolating Egypt's 50,000-man Third Army. The tide had turned.

As the Israelis retook the military offensive, the Arab world jarred from its feeling of euphoria. The sweet taste of victory, which had proved so elusive for twenty-five years, had again vanished. Even more disturbing was the continued support provided to Israel by the United States. It was, therefore, necessary for the Arab armies to resort to the strategically devastating fallback position of the October 1973 War: choking off the supply of oil, the energy lifeline of the West.

From 1953 until 1973 the price of a barrel of oil had risen by just one dollar from $2.00 to $3.00. But in just one day, October 16, the oil-producing members of the OPEC cartel raised the price of oil over 70 percent, from $3.01 to $5.12. Not content with this move, the Arab members of OPEC made another startling announcement on October 17: Oil production would be unilaterally cut by 5 percent a month until Israel withdrew from all the occupied territories. And as traumatic as that action was, Saudi Arabia, the supplier of over 20 percent of the free world's energy needs, went one step further the next day. It ordered a 10 percent cut in its oil production. Though the West barely had time to digest the full implications of each Arab action, massive Western panic set in immediately. In what amounted to a casus belli, Saudi Arabia and other Arab oil producers announced that all shipment of their oil to the United States would be embargoed.

Unable to enforce the embargo by itself, Saudi Arabia, which had been providing the United States with most of its imported oil, demanded that the oil companies enforce it. Aramco officials fully complied, diverting all shipments of oil from the United States. They later claimed that the Saudis had, in effect, put a gun to their heads. They gave customs information, shipping data, and consumption patterns to Sheikh Yamani. Despite the crippling effects of the embargo, the Saudis attempted to expand it; they asked for and received from Aramco all data relating to direct and indirect shipments of oil to American military bases and installations throughout the world. The fact that Aramco so willingly handed over such sensitive data later proved to be embarrassing in testimony by oil officials before an investigating Senate subcommittee.

On the battlefield in the Middle East the warring armies had agreed to a cease-fire. But the West was still numb. Just as American intelligence was caught entirely off guard by the Egyptian and Syrian attacks, so too were they blind-sided by the even more catastrophic decisions of the Arab oil producers. All over Washington policymakers, officials, and elected representatives attempted to figure out how to cope with the sudden international upheaval, which disrupted the relative tranquillity of the previous twenty years.

As gas station lines materialized in the United States for the first time since World War II, deep resentment and anger became obvious among the American people. A very cold winter made matters even worse. The economy spiraled downward like a plane out of control, putting hundreds of thousands of people out of work. In the fertile prairies of the rural Midwest, farmers demanded a food embargo on the Arabs. President Gerald Ford, echoing a popular sentiment even after the embargo had ended, intimated that the United States might resort to military action. "Throughout history," he warned, "nations have gone to war over natural advantages such as water or food." Even bankers in Western Europe, as reported by *Business Week,* were privately urging the United States to use military force to bring down the price of oil.

But in the Dhahran, New York, and San Francisco boardrooms of Aramco, Exxon, Mobil, Texaco, and Standard Oil, company officials breathed a sigh of relief. Sure, the American press had given the oil companies a hard time for displaying loyalty to profits rather than to the national interest. But where it really counted, in the king's palace in Riyadh and in their pocketbooks, they had done pretty well. So well, in fact, that on October 25—only five days after the embargo was announced and only two days after the battles stopped raging in the Sinai desert—Aramco's president Frank Jungers received wonderful news. A key Saudi official told him that Saudi Arabia was quite pleased by Aramco's pro-Arab stand. When Jungers remarked that the oil "cutback had not been good for our [Aramco's] business," the Saudi official promised him that Aramco would be rewarded.

Despite the embargo, the third and fourth quarters of 1973 proved to be one of the most profitable periods for the oil companies. They were able to sell to other countries the crude originally destined for the United States at a much higher price. Standard Oil, for example, posted a 54 percent increase over its 1972 profits. Exxon's profits registered a 59 percent increase over the previous year. The company's $2.5 billion in profits represented an "all-time record for any corporation, anywhere [and] at any time."[7] With flying colors, the American oil companies had passed Faisal's loyalty test.

3

THE PETRODOLLAR "REVOLUTION"

There is no money restraint. Just manpower, organization and supplies.

*American consultant
on the occasion of his
company's receiving
a contract to advise
Saudi Arabia on
industrialization*

American businessmen would also pass a loyalty test. For two weeks, beginning January 17, 1975, forty-five of America's top corporate leaders traveled through the Middle East, traversing 16,300 miles, on a special tour sponsored and organized by *Time* magazine. The business executives—whose firms in 1974 had sales of $100 billion and employed more than 1.5 million people—and *Time*'s editors met with eight heads of state, including President Anwar el-Sadat of Egypt, Prime Minister Yitzhak Rabin of Israel, King Hussein of Jordan, the emir of Kuwait, the crown prince of the United Arab Emirates, and King Faisal of Saudi Arabia. A round of talks in Beirut with the representatives of the Palestine Liberation Organization was also arranged.

Included in the U.S. corporate contingent were Gerald Trautman, chairman of the Greyhound Corporation; Henry Heinz II, chairman of the Heinz Company; Edward Carlson, chairman of United Airlines; William Hewitt, chairman of John Deere and Company; Andrew McNally IV, president of Rand McNally and Company; and Harry Henshel, chairman of Bulova Watch. The tour, reported *Time* magazine, "provided a unique opportunity to learn firsthand about a geopolitically vital region, and to pose hard questions to heads of state on oil and investment policy, petrodollar recycling and the prospects for war or peace."[1] The opportunity, however, "to pose hard questions" to King Faisal failed to materialize, according to both a *Time* report and a participant, as the king delivered a long, somewhat rambling discourse on the "evils" of Zionism and on the need to strengthen U.S.-Saudi ties.

One of those who returned to the United States with some strongly held beliefs was Robert H. Malott, the forty-nine-year-old chief executive officer of the Chicago-based FMC Corporation. A diversified manufacturer of such goods as agricultural machinery, mining equipment, industrial chemicals, and tracked military vehicles, FMC had worldwide sales of over $2 billion in 1974 from its operations in eighteen foreign countries and the United States.

On his return, Malott addressed the annual banquet of the Chamber of Commerce of Lawrence, Kansas. Speaking to an audience of six hundred in March 1975, Malott urged the United States to adopt an "evenhanded" policy in the Middle East, asserting that American policy had been "biased" in favor of Israel. He also relayed to the crowd his and his co-travelers' personal impressions of the various leaders. Syria's President Assad was a "pleasant surprise"; President Sadat was "very relaxed and very sincere"; Jordan's King Hussein was a "real gentleman"; and Saudi Arabia's King Faisal was " very open, very friendly." Malott heaped lavish praise on the king, adding, "[W]e were all very impressed with his concern and support of the United States." Malott even portrayed his meeting with the PLO supporters sympathetically. As for the Israelis, they displayed more "intransigence" than he and colleagues had expected.[2]

Within the next ten years, FMC, under Malott's stewardship, would receive more than $600 million in business orders (from armored personnel carriers to cargo-handling equipment to food-processing equipment) from Saudi Arabia, Kuwait, and other Arab countries. Soon FMC began to print part of its annual corporate report in Arabic. Moreover, several years down the road, Malott would play a pivotal role in the sale of sophisticated arms to Saudi Arabia.

Malott's comments in 1975 represented the beginning of the American business community's "discovery" of Saudi Arabia and other Arab oil-producing nations—and with good reason. The second-stage effects of the so-called 1973 "oil price revolution" were just beginning to be felt throughout the world. The real "revolution" had very little to do with oil. It concerned money—hundreds of billions of dollars. The entire international financial system was facing its most dire, potentially catastrophic challenge since the rules and institutions governing it—such as the International Monetary Fund—were created in 1944 at Bretton Woods, New Hampshire.

Some Arab countries were earning 12.7 million dollars an hour; in 1974, that was the rate at which revenues from the sale of oil flowed into the treasuries of the OPEC oil producers. By the end of that year, eleven countries had earned $117.8 billion dollars. These eleven are the principal oil producers in OPEC: Algeria, Indonesia, Iran, Iraq, Kuwait, Libya, Nigeria, Qatar, Saudi Arabia, United Arab Emirates, and Venezuela. The year before the price hike, 1972, these same eleven OPEC countries had collected $24 billion—just one-fifth of their 1974 revenues.

By the end of 1978, OPEC had earned a staggering $603.5 billion, a sum that constituted about two-thirds of the 1978 value of the total Eurocurrency market, which equaled $930 billion. Of that $603.5 billion, the lion's share—$474.6 billion—went into the coffers of just eight Middle East countries: Algeria, Iran, Iraq, Kuwait, Libya, Qatar, Saudi Arabia, and the United Arab Emirates. The flow of dollars from the oil-consuming countries to the oil-producing countries represented the most massive transfer of global wealth ever—far surpassing even the gold taken from the Americas by Spain five hundred years ago.

At first it seemed as if the international banking system might actually collapse under the weight of the one-way flow of money. Prominent bankers and financial analysts openly expressed their fears that the international financial system could not withstand the loss. In July 1974, for example, CIA analysts warned that Arab investments in the Eurodollar market were already reaching "the limits of prudent expansion" and that the danger of a worldwide collapse was ominous.[3]

These fears were not realized thanks to "petrodollar recycling," the process by which oil revenues collected by the oil producers were cycled back to the economies of the oil consumers. Oil producers invested their wealth in stocks, bonds, financial securities, and bank deposits (the latter allowed petrodollars to be reloaned by the multinational banks to poorer countries to finance their exorbitant oil debts), and purchased technology and imports for massive modernization programs. In effect, then, the money that the typical American motorist paid at the gas pump would come back to the United States in the form of petrodollar investments or the purchase of American goods and services.

As the producer of one out of every three barrels of OPEC oil, Saudi Arabia emerged as not only the most powerful oil country but the wealthiest as well. By the end of 1975, Saudi Arabia's position as holder of foreign reserves shot up to number two in the world, surpassed only by West Germany.

Impelled by the urgent need to sop up the surplus petrodollars, the United States created the Joint Economic Commission with Saudi Arabia in June 1974. The commission was officially designed to help build a Saudi bureaucracy and to promote industrialization in the kingdom. In essence, though, the real reason for its creation was to cement long-term ties between the two countries and to ensure that Saudi Arabia would spend its newfound wealth in the United States.

Numerous agreements were signed that called for U.S. personnel from the entire spectrum of the American bureaucracy—Food and Drug Administration, Federal Highway Administration, National Science Foundation, Department of Transportation, and Farm Credit Administration, to select a random few—to help the Saudis set up their own governmental bureaucracies. In the next nine years, nearly three-quarters of a billion dollars would be spent on such projects as training Saudis to be customs officials, traffic policemen, accountants, auditors, farmers, census takers, computer operators, and even space scientists. (Congress later discovered that the salaries of the 1,500 American officials working in Saudi Arabia or in the United States for the Joint Economic Commission were being paid by the Saudi government.) In order to accommodate Saudi needs, a separate office within the Treasury Department—the office of Saudi Arabian Affairs—was established, the only such office ever created for any foreign country.

And so, in 1974, Saudi Arabia—coaxed by American planners who saw in Saudi Arabia a social-urban laboratory that represented the opportunity of a lifetime, lured by multinational corporate entrepreneurs who offered hefty commissions, and won over by American government officials who promised miraculous results—launched a colossal five-year development project. Known as the Second Five-Year Plan, it was designed to transform the arid, primitive, mostly illiterate desert kingdom into a modern industrial state. The price—$142 billion, which later rose to $180 billion as a result of inflation and cost overruns. Weighing over 200 pounds, the plan—thousands of pages long—was every bit as ambitious and 3½ times more expensive than the Apollo Space Program, which had launched eighteen rockets into space and achieved six landings on the moon.

An entire national economic and social infrastructure was to be created virtually from scratch in a country with a population of seven million people spread out across 800,000 square miles, roughly the size of the United States east of the Mississippi. Saudi Arabia would buy everything that a nation with unlimited resources could afford: highways, airports, new cities, hotels, apartment complexes, power generating plants, shopping centers, offices, refineries, ports, a new air force, a modernized army, desalination plants, hospitals, plus every possible consumer good manufactured or made to order, ranging from yachts to refrigerators. These plans called for spending $8 billion on housing, which included construction of 175,000 new homes (to accommodate a full 10 percent of the population); $9 billion in educational construction, including 2,000 schools; $13 billion for a vast sewer and drainage system; and $3 billion for airports, 8,500 miles of paved roads, and 2,000 miles of electrical power transmission lines.[4]

One of the first American advisers on the scene was the California-based SRI International. At one point it had been connected to Stanford University, but later it became an independent research and strategic planning corporation providing specialized consulting for a vast array of clients, including the U.S. Department of Defense, multinational corporations, and foreign governments. Its projects ranged from evaluating American strategic doctrine and weapon systems to predicting trade flows from Cuba in the event that the American economic quarantine was lifted. On its board of directors and its international council have sat some of the most prominent figures in American corporate and government life such as A. W. Clausen, then president and chief executive officer of the Bank of America (and now head of the World Bank); Albert Casey, chairman and president of American Airlines; William W. Boeschenstein, president and chief executive officer of Owens-Corning Fiberglass Corporation; George P. Shultz, president of Bechtel Corporation (later to become U.S. secretary of state); and dozens of others.

Though SRI had provided assistance to the Saudis since 1968 in devising the first Five-Year (1970–75) Development Program, which had a budget of only $9.2 billion, the real challenge came in 1974. SRI compiled a 579-page document, using fifty full-time SRI employees including petroleum economists, civil engineers, health specialists, and urban economists.[5] Over 500,000 technicians, laborers, and managers were to be imported from other countries to help implement the program, which Saudi Arabia adopted as part of its second five-year plan. "There is no money restraint," declared Wilson Harwood, SRI's Middle East representative, in August 1975, "just manpower, organization, and supplies," on the occasion of SRI's signing of a $6.4 million contract with Saudi Arabia to continue to provide research and consulting to the Saudi government.[6]

As the Arab states prospered, so did SRI. By 1977 the volume of SRI's Middle East and Southeast Asian contracts grew by a whopping 45 percent over the previous year as Kuwait and the United Arab Emirates enlisted the company's help in spending their money.[7] Moreover, SRI's contract with the Saudis spawned second-tier consulting contracts in which the Palo Alto firm advised American and foreign companies on how to solicit Saudi and other Arab business.

But the value of SRI's contracts was small compared to the mammoth construction and engineering projects that began to be awarded to American multinational firms. In August 1974, the San Francisco–based Bechtel Corporation, a privately held firm whose revenues place it among the nation's largest concerns, landed the $3.4 billion project to build from scratch the ultra modern King Khalid International Airport in Riyadh.[8] When finally completed, the airport will extend over an area six times the size of Manhattan and be capable of serving 15 million passengers a year.

The King Khalid Airport was not Bechtel's first project in Saudi Arabia. In 1947 the firm was the contractor for the 1068-mile Trans Arabian Pipeline (TAP) that pumped Saudi oil from the Persian Gulf across the Arabian Peninsula through Jordan and Syria to the southern Lebanese port of Sidon on the Mediterranean, where it was shipped to Europe. In the opinion of some long time observers, close links between the CIA, State Department, and Bechtel officers in the 1950s and 1960s made the company a de facto arm of the American government in the Middle East. Besides serving as American eyes and ears, Bechtel officials routinely transmitted messages between American and Middle East leaders, and also ensured that American "interests" in the region were protected. Two ex–CIA directors, John A. McCone and Richard Helms, and one former ambassador to Saudi Arabia, Parker Hart, actually went to work for Bechtel when they left the government.

Bechtel assigned eight hundred of its employees to work full-time in the United States on Saudi projects, making that country the firm's largest source of foreign revenue in the 1970s. In 1975 Bechtel was selected to supervise the construction of the $9 billion petrochemical complex and city of Jubail.

Located on the Persian Gulf forty miles north of Dhahran, Jubail was one of two small backwater Saudi villages that had been selected to undergo a transformation into a huge bustling industrial city. The final cost has been revised upward of $40 billion. Bechtel put over fifteen hundred architects, engineers, and other professionals to work on the Jubail project, along with a labor force of forty thousand workers.

The other village was Yanbu, nestled on the other side of the kingdom on the Red Sea, 230 miles north of Jidda. In the 1960s Jubail and Yanbu had a combined population of less than 15,000. By the year 2000 their populations are expected to zoom to 400,000. These cities form the cornerstone of Saudi Arabia's industrialization program, designed to reduce the country's future dependence on oil exports.

Jubail will soon become a sprawling industrial complex covering sixty-six square miles with four refineries, two $300-million fertilizer plants, a $600-million iron mill, and five petrochemical plants that together will cost a total of $4 billion alone. Jubail's twin city, Yanbu, will be equipped with similarly impressive petrochemical facilities.[9]

The Ralph M. Parsons Company, based in Pasadena, California, was selected to supervise the $20-billion construction of Yanbu after the Saudis had accepted the company's $6.5-million master plan in 1975. American companies were also selected to build the 750-mile pipeline transporting liquefied natural gas and oil from Yanbu to the eastern provinces and the giant Persian Gulf port of Ras Tanura.

Aramco, in addition to producing 97 percent of the kingdom's oil, undertook in 1975 to build the $4.5 billion gas-gathering project designed to produce 6 billion cubic feet of gas a day—equivalent to the amount needed to fuel "half the gas-using homes" in the United States in 1970.[10] Aramco's expenditures in Saudi Arabia were so huge that it emerged as a separate purchasing arm of the Saudi government. Besides imports of 2,825 tons of heavy lift cranes, for example, its 1976 import bill of $740 million included such diverse products as 27,000 pounds of Oreo cookies and 208,000 cans of potato chips.[11] By 1977, the oil consortium's purchases reached $2.6 billion—an amount that surpassed the national budgets of over seventy-five countries at that time.

Like the endless concentric rings produced by a stone thrown into a lake, the vast ripple effect of the petrodollar boom was felt by companies that had no direct dealings with the Saudis whatsoever. Across the United States, petrodollars filtered intact through the many subsets of the economy. In the economically depressed town of Jeannette, Pennsylvania, a half-hour east of Pittsburgh, the Elliot Company received a $50-million order from Saudi Arabia in 1976. The contract was for fifty-two 100-ton gas compressors needed for the Saudi gas-gathering program. Elliot figured prominently in the economy of Jeannette (population 15,000), providing 40 percent of the town's industrial jobs; the Saudi orders enabled the company to hire an additional 554 workers. Twenty-five miles to the southwest in the small Pennsylvania town of Donora, Elliot established a 125-man lubrication division that mounted the compressors for shipment to Saudi Arabia. And supplying additional parts for Elliot's compressors were scores of Pennsylvania's small businesses. Moreover, the Elliot contract provided an economic boost to companies as far away as Minden, Nevada, where Bently Nevada assembled electronic sensors for the compressors, and Foxboro, Massachusetts, where the Foxboro Company produced temperature- and pressure-control devices.[12]

The Saudi market rescued some firms from the doldrums of the OPEC-induced recession that had gripped the country, propelling them into the economic stratosphere. J. A. Jones

Company, a construction firm based in Charlotte, North Carolina, for example, had not done any substantial overseas work since the 1930s. But in August 1974 the company set up shop in Saudi Arabia and won the first contract it bid on. A year and a half later the firm had received contracts for projects totaling $385 million from the Saudi government, Aramco, and the U.S. Army Corps of Engineers. Its work force in the kingdom exceeded 2,600.

Interviewed in *Aramco World* magazine, George Turner, vice-president of J. A. Jones, declared, "We now have in the kingdom more assets than we've ever put into a country outside the United States."[13] Moreover, he added, the firm had imported over 100 million dollars' worth of goods from companies located in thirty-seven of the fifty states. The multiplier effect of J. A. Jones's Saudi contract was vast. Added Turner, "Other than some locally produced cement and reinforcing steel, everything it takes to build a building is imported—everything, and we import the bulk of it from the United States. All the electrical equipment from heavy switch gear and transformers down to cables, receptacles, and light switches. We even import furniture and linens from the U.S."

CRS Design Associates of Houston experienced a similar success story. As described by Tom Curtis, in the *Texas Monthly,* most architectural firms in the United States in 1974 and 1975 were feeling the brunt of the oil-induced recession. Construction activity had come to a virtual standstill, and many firms and architects were out of business, let go, or bankrupt. But not CRS; its business was booming. By 1977 the firm had been given so much Saudi business that it had become "the nation's largest grossing architectural firm," employing 700 employees—a giant leap forward from a few years before when it had only 250 workers.[14]

The first CRS project in Saudi Arabia began in the late 1960s when the company was awarded a $50-million contract—paid for primarily by Aramco—to design the University of Petroleum and Minerals in Dhahran. Favorably impressed, the Saudis threw hundreds of millions of dollars' worth of business to CRS in the oil boom years. By 1976 CRS had been awarded multiple contracts to design roads, housing, and schools—such as the $3.4 billion University of Riyadh, which CRS designed in association with four other American architectural firms. By some estimates, the total value of its projects in Saudi Arabia and two other Persian Gulf nations, Bahrain and Kuwait, amounted to more than $6 billion. So it was not surprising that by 1976, 59 percent of CRS's pre-tax earnings came from foreign-based projects, most of them in the Middle East. By 1977, half of the firm's $34 million gross revenue came from the Middle East. The following year, Saudi financier Ghaith Pharoan purchased a 20 percent interest in the firm for $5.5 million.[15]

For military contractors the Middle East appetite seemed almost insatiable. Increasing the kingdom's military apparatus became a national obsession. Saudi Arabia's defense budget zoomed from $2.8 billion in 1973 to $10.3 billion in 1978. And the United States allowed—indeed encouraged—the kingdom to purchase military equipment way in excess of the country's ability to absorb it. Though the unlimited arms sales were rationalized as an extension of the Nixon administration's "twin pillar of stability" doctrine—which relied on Saudi Arabia and Iran to act as surrogate American policemen in the Persian Gulf—the real reason related to the need to recycle petrodollars. Many Saudi officials and sales agents—enticed by the lure of multimillion-dollar bribes and commissions by American defense manufacturers, mesmerized by the latest electronic gadgetry in presentations by Department of Defense officials who wanted to reduce the per unit costs of their own weapons, and driven by an intense nationalistic rivalry with Iran to become the singular Persian Gulf "superpower"—went on a virtually uncontrollable military buying spree.

The defense budget of Saudi Arabia, with armed forces totaling only 47,000, skyrocketed to become the sixth largest in the world by 1980, surpassed only by the United States, the Soviet Union, Britain, Germany, and China. Iran, too, under the megalomaniacal leadership of Shah Mohammed Reza Pahlavi who was bent on achieving Persian Gulf hegemony, also began importing billions of dollars' worth of tanks, fighter aircraft, and other military equipment. In fact, between 1972 and 1978, the United States sold to Iran over 14 billion dollars' worth of military hardware, construction, and services, a destabilizing development that later proved fatal to the shah's regime. Not since the European Allies began to rearm themselves in the late 1940s and early 1950s had American defense manufacturers experienced such a boom. Between 1974 and 1978, American defense manufacturers and contractors sold military materials worth more than $15 billion to Saudi Arabia alone. And by 1984 Saudi Arabia had become the largest recipient ever of American military sales, having spent a record $42 billion in just ten years.

Litton Industries, for example, which most Americans now associate with cooking devices such as microwave ovens, was given a $1.6 billion contract to provide a national military communication system for the Saudi air command and control system. HBH Company—a consortium of the Hughes Aircraft Company, the Bendix Corporation, and the California-based Holmes and Narver—was awarded a $671.2-million three-year contract to expand the operations of the Royal Saudi Navy. The Nebraska-based Vinnel Corporation was given a $77 million contract to train the Saudi national guard. A planned billion-dollar military city, named after Saudi King Khalid, was contracted out to the U.S. Army Corps of Engineers, which subcontracted the project to a group of American firms that included Morrison Knudsen of Boise, Idaho.

A $260-million contract was given to Raytheon for the Saudi Hawk missile air defense program. And in December 1975, Northrop received a $1.8-billion contract to provide manpower training and ground maintenance support for the F-5 aircraft that Northrup had sold to Saudi Arabia the year before for $750 million.

Americans continued to flock to Saudi Arabia and other Arab oil-producing countries in unprecedented numbers. Pan Am registered an increase of 19.3 percent in its 1974 air traffic to the Middle East over the previous year.[16] Saudia, the Saudi Arabian airline, experienced growth rates as high as 200 percent for passenger and freight volume, and opened new offices in a half-dozen American cities. It was not unheard of for large multinational construction firms to shell out $5 million annually just on airfare, ferrying their executives, employees, and laborers to the Middle East.

Established and powerful trade organizations, like the 12,500-member National Association of Manufacturers (NAM), sponsored industrial missions to Arab countries for scores of chief executive officers. As early as April 1974, NAM had set up a special task force—under the impetus of A. R. Marusi, chairman of Borden, Inc.—to study new ways of recycling petrodollars. Even state legislatures became involved: North Dakota, for example, passed a resolution promoting trade with Saudi Arabia and promptly authorized state trade missions.

Daily announcements of mega-contracts became routine for all types of companies. Waste Management, Inc., an Illinois-based firm, received a $243-million sanitation service contract for the city of Riyadh. The Minnesota-based Pillsbury Company won a multimillion-dollar contract to build flour mills in Saudi Arabia. General Motors was given the nod to build an assembly plant to manufacture eight thousand Chevrolet cars and trucks a year for sale in Saudi Arabia. A branch of Pan American Airlines was awarded a $100-million contract to build hotels in Riyadh and other Persian Gulf cities; Intercontinental, Sheraton, Holiday Inn, Hyatt, Hilton, and Marriott were also able to land business in the hundreds of millions. Turner Construction won a $30-million contract to expand a power station in Riyadh. Westinghouse Electric, after receiving a $45-million contract to install four turbine generators, mounted its most expensive trade show in history at a cost of $1 million.[17] Even United Press International hopped on the petrodollar gravy train: the news agency became a paid consultant to a Norwegian company hired to re-haul, expand, and reorganize the Saudi Arabian News Agency, an official press arm of the Saudi government.

Nor was the financial connection limited to imports of American goods and services. Top American corporations were able to tap into the huge surplus of Saudi oil revenues—tens of billions of dollars that the kingdom was amassing over and beyond its huge expenditures in its domestic economy. AT&T borrowed $650 million from the Saudi Arabian Monetary Agency, representing 1.7 percent of the company's total debt. IBM received a loan of $300 million, U.S. Steel got $200 million, and Chrysler received $100 million from eighteen Arab banks. In the next few years a total of $13 billion in Saudi funds would be loaned to American firms.

In 1974 and 1975 the American economy was mired in a deep recession, the severity of which was exceeded only by the Great Depression of the 1930s. The 600 percent increase in the price of oil was translated into an inflation rate of 10 percent, a post–World War II high; an unemployment rate of 8.5 percent involving 8 million people, the highest since 1941; a major drop in the GNP; and an oil import bill that shot up more than 300 percent, from $7.0 billion in 1973 to $23.4 billion in 1974. Industrial production fell for twenty-one consecutive months, accounting for a total drop of 13 percent.

The era of unlimited economic growth—a major component of the American dream—had suddenly come to a halt. The Saudi and other Arab spending sprees, in other words, could not have come at a more propitious time. Ironically, it was these very same countries that had caused the American economic crisis. From 1974 through 1978, Arab countries would constitute the fastest-growing trade market for American exporters. The Saudi market alone grew 1,000 percent, from $442 million in U.S. exports in 1973 to $4.8 billion in 1978. Two hundred American companies opened offices in Saudi cities; thirty thousand Americans lived there; and the growth in Saudi business provided employment to an additional 300,000 Americans in the United States.

But Saudi industrialization did not proceed without a political agenda for the kingdom's business partners. As demonstrated by King Faisal's success in eliciting the cooperation of Aramco executives in trying to change American policy, Saudi Arabia's business "friends" were expected to demonstrate allegiance. And the spring of 1975 provided a critical opportunity for the Saudis to "educate" the American business community and public.

4

AN
UNSENTIMENTAL
EDUCATION

Just consider that this year Saudi Arabia
is expected to pass the United States in
the size of its monetary reserves and that
within six years, the Middle East nations
will probably control three-fourths of the
world's monetary reserves. The market is
ripe for American ingenuity and know-how.

*From a mass mailing
sent to American business
executives across the
country inviting them to
attend economic seminars
hosted by visiting Arab
officials in early 1975*

On the eve of the unveiling of its dazzling $142-billion Five-Year Development Plan, Saudi Arabia dispatched to the United States seven officials in March 1975 on what was supposed to be an economic mission. Criss-crossing the entire country at a dizzying pace—ninety cities in just three months—the official Saudi delegation met with more than twenty thousand American business leaders. Investors, bankers, manufacturers, grain brokers, electrical contractors, and executives representing the entire spectrum of American industry and business clamored to hear the Saudi speakers.

In their presentations, the delegates focused extensively on the vast new export market emerging in Saudi Arabia. They offered attractive, almost irresistible tax incentives to American companies willing to establish subsidiaries in the kingdom. For example, American companies were told they would be granted a five-year moratorium on income tax, free land, and fifteen-year interest-free loans. Dangling the opportunities that entrepreneurs dream of, the delegates were, not unpredictably, besieged by Americans wishing to get a piece of the action. Seminars were jammed with standing-room-only crowds, the delegates were deluged with phone calls twenty-four hours a day. The mission's leader, Dr. Abdelrahman al-Zamel, a top Saudi education official, who later rose to a senior position in the Saudi cabinet, acknowledged to the *New York Times* that he was "astounded by the change in attitude" of American businessmen since his previous visit to the United States in 1973.[1] Unlike his earlier visit, al-Zamel was inundated by Americans clamoring to learn more about the Arab world.

But before the rapt American audiences around the country, the Saudi mission also insured that no one left his or her seat without being made fully aware of the inseparable link between politics and business. The Arab League's boycott of Israel, though in existence since the creation of Israel in 1948, was unknown to many Americans. In his speeches and press conference, al-Zamel described the Arab boycott as a "political instrument," no different from the American embargo on Cuba.[2]

For the mainstream of the American business community, which had never previously dealt with the Arab world—save for a select few American multinationals—politics and trade had always been kept separate. When farm machinery was sold to the Russians, there was no link to the Soviets' bitterly hated and feared rival, the Chinese. The Indians did not boycott their military and political adversary, Pakistan.

And even when foreign countries were engaged in boycotts of other nations, they had not demanded third-party adherence to sanctions against that nation. Black African states did not insist on foreign businessmen adhering to their sanctions against Portugal or against the apartheid regime of South Africa. Taiwan did not demand that trading partners honor its boycott of the Soviet bloc nations and China. The tens of thousands of American businessmen, investors, industrialists, and entrepreneurs, aspiring to forge a link to the petrodollar recycling process, were soon to discover just how different the petrodollar was from the ruble, the rupee, the drachma, or the franc.

Capping the three-month tour at a seminar attended by 250 business leaders on the morning of May 27, 1975, at the St. Regis–Sheraton in New York City, al-Zamel—who had completed his doctoral dissertation in 1972 at the University of California, analyzing the ineffectiveness of Arab public relations in the United States—not only defended the Arab boycott of Israel but also pinned blame on the United States for any future Saudi oil embargo: "Any new embargo would be a reaction to United States behavior. We know that Israel cannot go to war without the approval of the United States."[3] If his American listeners had no political opinions on the Middle East before the seminars, they most certainly were developing them; their cognizance of Arab political interests was now becoming an integral part of their economic horizon.

Simultaneously, the American business community was given an even more intensive lesson in politics by another Arab economic delegation trotting around the country. An extensive publicity campaign promoting seminars on Arab Oil and Investment Opportunities in the Middle East, in part sponsored by the semiofficial Beirut-based Arab Press Service, included an advertisement in the *Wall Street Journal,* and a mass mailing sent to thousands of American executives tantalizing them with this "fact": "Just consider that this year [1975] Saudi Arabia is expected to pass the United States in the size of its monetary reserves and that within six years the Middle East nations will probably control three-fourths of the world's monetary reserves. The market is ripe for American ingenuity and knowhow."[4]

In order to "foster candid discussion and because of the sensitiveness of some of the subject matter," the Arab Press Service promised to prohibit tape recorders and bar media participation. Led by Farouk Al Akhdar, a key adviser to the Saudi Arabia Central Planning Authority, the delegation traveled from coast to coast explaining "investment opportunities" to packed houses of businessmen, each paying a $185 registration fee. The Saudi official also rallied support for the Arab cause.

Indeed, as noted by the Arab Press Service, the primary purpose of Akhdar and his colleagues may have been political: "Six senior officials representing the Saudi government are currently touring the United States and serving Washington with a firm notice, subtly worded, but nevertheless strong enough to alert American officials and public opinion that the present 'no war, no peace' situation is unacceptable and will no longer be tolerated!''[5]

Their message was anything but subtle. In San Francisco, for example, where Akhdar was the guest of the World Affairs Council of Northern California, and in San Jose, where he was sponsored by the Santa Clara Valley World Trade Club, the Saudi official accused the ''Zionist movements''—by which he meant American supporters of Israel—of being responsible for the ''badly damaged Arab image in the United States.''[6]

Later, 2,700 miles to the southeast in New Orleans, Akhdar met with the city's Chamber of Commerce and warned that his country was limiting its investment abroad because of alleged ''Zionist'' pressure on Congress to stop Arab investment in the United States.[7] Actually, Congress considered legislation restricting Arab investment if such investment elicited American corporate sanctions against American Jews or American companies. However, no legislation curbing Arab investment has ever been passed.

At the elegant Plaza Hotel on May 2 in New York City, Akhdar wound up his tour with the blunt warning—though he said he was expressing his ''personal views''—that Saudi Arabia would be reluctant to invest its money in ''any country that takes an antagonistic policy against the Arabs, against Saudi Arabia.''[8] Like al-Zamel, Akhdar vowed that the Arab boycott of Israel would continue. He compared it to the American embargo of Cuba in an analogy that was fast becoming a common motif in the Arab lecture to the West.

In reality, as the American business community was about to discover, the Arab boycott of Israel was vastly different from the American embargo on Cuba. Under the terms of their boycott, the Arab governments refused to trade with Israel just as the United States refused to trade with Cuba. To ensure that Israeli goods did not enter the country through a third port, the

Arab nations demanded certification that exports to their countries had no Israeli components. And shipping vessels that stopped at Cuban ports were barred from docking at American ports. But there the similarity ended. The Arabs also banned commerce with all firms that had economic dealings with Israel and firms that had Jews or known "Zionists" in prominent positions. Also subject to the blacklist were companies that would not respond to Arab requests for information about their relationship with Israel. Moreover, some American firms were on the Arab blacklist simply because they traded with other American firms that did business with Israel. There was no other boycott as extensive or as multi-tiered in the world.

In the capital of Syria, Damascus, a Central Arab Boycott office administers the boycott by collecting information from mandatory questionnaires sent to firms and from publicly available data. From this information the officer assembles a master worldwide blacklist of firms. In addition, other Arab nations keep their own independently compiled lists. RCA, Zenith, Coca-Cola, Bantam Books, Sears Roebuck, and Revlon have been among the hundreds of American firms placed on the blacklist. So have actors and singers and the motion pictures in which they appear—among them Harry Belafonte, Helen Hayes, Kirk Douglas, Jerry Lewis, Paul Newman, Sophia Loren, Elizabeth Taylor, Frank Sinatra, and Phil Silvers. Even the Walt Disney movie *Snow White and the Seven Dwarfs* was blacklisted because, in it, the prince's horse is named Samson, after a hero in Jewish history.[9]

Though these two cross-country tours were pivotal events in the political mobilization of the American business community, they were not the first. Indeed, efforts to subtly mingle politics with business had begun the previous year. One of the first such programs to induce corporate support for the Arab position was unusual because the American government was involved. In mid-November 1974—just one year after the devastating oil embargo—invitations were sent to thousands of northeastern business executives by the U.S. Department of

Commerce to avail themselves of the economic opportunities of "unprecedented proportions" by participating in a Commerce Department–sponsored seminar on "doing business in the Arab world." This seminar was just one of nine that the Commerce Department arranged throughout the country in the latter part of 1974 to encourage American business to export to the Middle East.

On December 4, 1974, over three hundred businessmen jammed into the Terrace Room of the Plaza Hotel in New York to hear what they were told would be presentations on the "business climate, sales prospects, and development projects" in the Arab world.[10] But when Lucius D. Battle, the former American ambassador to Egypt (1964–67) and former assistant secretary of state in the State Department's Near East Bureau, delivered the keynote address, he sounded a political note akin to the message given by the Arab economic delegations that would tour the United States five months later.

Lambasting the "utter rot in our press" for its "sanctimonious and absolutely emotional" criticism of the Arab oil embargo, Battle, like the Arab officials, compared the embargo to the American trade sanctions against Cuba.[11] Battle—who at the time was head of the Middle East Institute, which received more than 40 percent of its 1974 funding from oil companies—claimed that the impact of the oil price hike was no less severe to "many countries" than the "rise in the price of food, wheat, and the difficulty of obtaining fertilizer," an assertion simply not based on fact.[12] Battle himself gave other speeches during this time in defense of Saudi Arabia.[13]

Through seminars like these in America and in Saudi Arabia, and through direct prodding by Arab government officials stationed in the United States, the American business community was becoming acutely aware of the connection between politics and trade.

Initially, the major issue confronting American business officials was the Arab boycott. To many companies the boycott presented a Hobson's choice: either they could stand on principle and lose important business opportunities, or they could accede to the boycott. With billions of dollars at stake, their choice, not surprisingly, was in favor of the boycott.

Commerce Department records revealed that among the companies that complied with the Arab boycott of Israel at various times and degrees were General Electric, John Deere (farm machinery), White-Westinghouse (appliances), Jim Beam Distilling Corporation, Minnesota Mining and Manufacturing Co., and Rockwell International (aircraft and military weapons). According to Commerce Department officials, the boycott instilled such fear that even when compliance was not required of them, firms responded with statements to Arab officials avowing their support of the boycott. The effect was frightening. Yet there was no legal means of stopping the pervasive intimidation of American business.

Some firms looked to the American government for guidance—but no action or response was forthcoming. If anything, the Ford administration implicitly encouraged compliance with the boycott. The U.S. Treasury Department actually released guidelines to American businesses in late 1976 informing them how to evade some of the mild anti-boycott tax penalties that Congress had just passed. Senator Abraham Ribicoff of Connecticut dashed off a letter to Treasury Secretary William E. Simon during the first week of December accusing the Treasury Department of "encouraging circumvention" of the newly approved congressional statutes.[14]

Soon, American firms by the hundreds began complying with the sanctions against Israel, and against the tertiary targets of the boycott—other American companies. In one six-month period between March and September 1976, for example, 94 percent of the 11,000 commercial transactions reported to the Commerce Department were found to have upheld various levels of the boycott demands.[15]

Some companies, however, appeared to follow the letter and spirit of the boycott just a bit too zealously. Bechtel's practices were considered by the Justice Department to be so flagrant that it sued the company in 1976 for "conspiring" with four affiliates in refusing to conduct business with firms blacklisted by the Arabs.[16] Moreover, Bechtel—which had subcontracts with five hundred firms on its Jubail project alone—went a step further than many other American companies in acquiescing to Arab demands. The Justice Department suit charged that Bechtel—above and beyond its own blacklisting of companies that traded with Israel—had also prohibited its subcontractors from dealing with blacklisted companies. In January 1977 Bechtel agreed to a consent decree—although it did not admit the allegation—that forced the company to end its blacklisting of American firms. Later the firm unsuccessfully sought to annul the agreement.

Prominent American banks—including Chase Manhattan, Morgan Guaranty, Bank of America, First National Bank of Chicago, and Bank of San Francisco—were said to have processed exporters' letters of credit that contained proof that a company had no commercial dealings with Israeli firms or with American and other firms on the Arab blacklist. Despite the fact that it handled Israeli bonds in the United States, Chase Manhattan refused in 1975—the same year it opened a branch office in Moscow—to open a branch office in Israel. Its chairman, David Rockefeller, left no doubt as to why. "If we were to open a branch there," he said, "all our business in the Arab world would come to an end."[17] American as well as European investment houses were asked to drop "Jewish" banks—that is, banks founded by Jews—that participated in underwriting international loans. For example, in February 1975, the Kuwait Investment Company demanded that Merrill Lynch Pierce Fenner & Smith drop the American affiliate of Lazard Freres from participating in underwriting a loan; but Merrill Lynch refused, and Kuwait withdrew from the syndicate. Arab governments, however, had been successful in blocking the participation of the investment houses of N. M. Rothschild and S. G. Warburg in syndicating other loans.[18]

In 1975, World Airways, the largest charter airline in the world, was awarded a contract to run charter flights for some of the two million Muslim worshipers on their annual pilgrimage to Mecca. Shortly thereafter, the airline faced a class action religious discrimination suit, in which it was charged with having asked its Middle East–bound employees for a "letter from a church showing membership, or proof of baptism or marriage in a church."[19]

One firm actually pressured some of its Saudi Arabia–bound American employees to convert to Islam in order to qualify for Saudi business. The Texas-based branch of Dynalectron Corporation, which has its headquarters in McLean, Virginia, recruited and trained pilots under a subcontract to a Japanese firm to operate advanced fire-fighting helicopters in the city of Mecca, the holiest shrine in Islam—so holy, in fact, that Saudi Arabia allows no one who is not Muslim to enter it. So, during the orientation program in Japan for its newly recruited American workers on their way to Saudi Arabia, Dynalectron arranged for its employees to convert to Islam, even hiring an Islamic consultant to facilitate the conversions. Moreover, those hired to work outside Mecca—and thus not subject to the ban on non-Muslims—were also asked to convert. In the end, thirty of the forty-five pilots converted.[20]

Following pressure by Arab officials, scores of companies and even universities began weeding Jews out of their projects and other contractual programs in Arab countries. Vinnel Corporation, a California company that trained the Saudi national guard, insisted that personnel would not be allowed into Saudi Arabia if they had any "contact or interest" in countries not recognized by Saudi Arabia. Whittaker Corporation, a Los Angeles–based hospital service and supply firm, was sued by an American employee for allegedly being fired for referring to Israel as the "Holy Land" in the company newsletter for which she wrote when she was stationed in Abu Dhabi.[21] The woman employee had worked for Whittaker for nineteen years. Baylor College of Medicine in Texas refused to send Jews to Saudi Arabia for its lucrative cardiovascular surgical contract with King Faisal Hospital. Other universities, however, such as M.I.T., the University of Colorado, and the University of Washington, to their credit, demanded that nondiscrimination clauses be included in their contracts. M.I.T. lost its contract after its president, Dr. Jerome Weisner, stipulated in a letter that religious discrimination would nullify the contract. Prince Mohammed Ibn Faisal found this to be a "threatening letter." [22]

One of the most shocking revelations occurred when congressional hearings in 1975 disclosed that the U.S. Army Corps of Engineers had acceded to Saudi demands that all American military personnel serving in Saudi Arabia as well as those retained by American companies on contract to the Corps of Engineers submit certificates of religious affiliation and other background material—actions that effectively barred Jews from participation on Saudi projects. American blacks, too, were not given military assignments in Saudi Arabia as a result of Saudi Arabia's insistence. [23]

In Congress, pressure mounted for federal legislation to curb discrimination against American Jews and to deal with the portions of the boycott that punished American firms trading with Israel. (The Arab boycott of Israel itself, however, was accepted as a legitimate prerogative of the Arab states, and neither Congress nor any presidential administration tried to curb it.) Finally, in November 1975, the Ford administration announced that it was implementing a series of steps to prevent government and corporate exclusion of, and discrimination against, Jews.

But there was no effort whatsoever to counter the Arab nations' enforcement of the boycott of American firms. In essence, the secondary and tertiary components of the boycott—for which Bechtel was cited—remained intact. By early 1976 it became apparent that the Ford administration was actively trying to dissuade Congress from introducing effective corrective legislation. Information about the boycott was denied to congressional committees, and Ford officials offered a series of legal palliatives, but there were no real teeth to any of them.

From the administration's vantage point, there was good reason for its actions. Officials of Saudi Arabia and other oil-producing nations had let it be known in meetings with Ford administration officials that any congressional or executive action designed to curb any part of the Arab boycott would be perceived as an infringement on their "sovereignty" and would be injurious to all American trade with the Arab world. This threat, however, would later turn out to be hollow.[24]

Nowhere was this message heard more loudly than in business circles, which launched a massive campaign to stop congressional passage of anti-boycott laws. "[I]t is up to U.S. firms with vital interests here [in the Middle East] to safeguard the right to work and earn. Others are willing and ready to take their place," said Hisham Nazer, Saudi Arabia's minister of planning.[25] National trade associations, such as the Associated General Contractors of America and the U.S. Chamber of Commerce, lobbied their legislators or testified before Congress against anti-boycott action. Mobil Oil took out a series of national advertisements in the *New York Times* and other newspapers, warning that anti-boycott legislation would transform the United States into a "second-rate economic power" and would adversely affect the American life-style.

And on the evening of the September 23 debate between presidential opponents Jimmy Carter and Gerald Ford, Mobil posed a series of questions in a prominently featured *New York Times* ad, including the following: "Are they [the candidates] willing to jeopardize American access to vitally needed foreign crude by supporting legislation which claims to protect American companies from the effects of the Arab boycotts, but which would actually foreclose to American companies the world's largest crude oil reserves and one of the world's fastest growing consumer markets?"[26]

Dresser Industries, a Dallas-based manufacturer of oil drilling equipment, bought a two-page spread on April 14, 1977 in the *Wall Street Journal* that predicted anti-boycott legislation would result in a loss of "500,000 jobs." Bechtel, Fluor, and Pullman Kellogg (all giant engineering and construction companies) employed a Washington public relations firm to coordinate anti-legislation activity by other American firms. Caterpillar Tractor, Ralph M. Parsons Company (construction), Exxon, Continental Oil, and other companies also joined the lobbying effort. William E. Leonhard, president of Ralph M. Parsons charged in an interview with *Business Week* that anti-boycott bills would "damage U.S. companies, the U.S. economy, and the country as a whole for the benefit of a select few."[27]

In late September 1976 the clarion call was sounded when the Saudi foreign minister, Prince Saud al-Faisal, spoke to fourteen hundred businessmen in Houston, telling them that Saudi Arabia would not countenance any anti-boycott legislation, and warning of adverse effects on American trade, economy, and energy.[28] Aramco's Houston offices took the unprecedented step of sending certified Mailgrams to four thousand of its suppliers across the United States warning that passage of any congressional legislation would "severely jeopardize, if not render impossible," continued American trade with the Arabs—which included Aramco's own plans to pay $30 billion to more than two thousand firms for products related to its own Saudi projects. Moreover, Aramco warned that legislation would even have deleterious repercussions on the peace negotiations and would jeopardize the security of American oil supplies.

Beyond influencing compliance with the boycott and lobbying against anti-boycott legislation, the petrodollar was beginning to make itself felt in various parts of the United States where spontaneous political gratitude was taking root. One of the most telling cases occurred in a small rural community in Arkansas.

In December 1975, Ward Industries, which manufactured school buses, received a $20-million contract to build seven hundred specially designed buses (to shuttle Muslim pilgrims) for Saudi Arabia. The contract was not large compared with the many multibillion-dollar contracts Saudi Arabia had signed, but the impact on Ward and on the community in which it was located was profound. For Ward, situated in Conway, Arkansas, was a town of only 15,510 people, located twenty-five miles north of Little Rock. An Egyptian order for 60 million dollars' worth of buses soon followed, and consequently over three hundred new employees were added to the assembly line. The Egyptian order alone accounted for 25 percent of Ward's annual production, which meant an additional 336,000 man-hours of work.[29] By mid-1977 plans were solidified for Ward to open an assembly plant in Jidda as a joint venture with Saudi businessman Ghaith Pharoan. In the course of the commercial transactions, Ward sent scores of its mechanics and other workers to Saudi Arabia to provide technical assistance (such as repairing the buses), and forty workers from Saudi Arabia spent several weeks in Conway being trained by Ward Industries.

Everyone in the small town of Conway soon became acutely aware of the Saudi connection. Interviewed for *Aramco World* in late 1976, Bill Ward, the advertising director for Ward Industries, proudly boasted: "I'll bet there isn't a town Conway's size anywhere that has a greater awareness and appreciation of Saudi Arabia."[30] The company officers who returned to Conway after visiting the Middle East became goodwill emissaries on behalf of Saudi Arabia. After having addressed the Lions Club, the Civic Club, and the Business Women's Club, where he explained his attitude toward Saudi Arabia, Vice-President O. P. Ryan asserted in an interview with the *Washington Post:* "We can't afford to make the Arabs mad at this time."[31]

Elsewhere, Arab officials concentrated on different social and racial groups. Saudi officials and investors openly courted American blacks. In 1977, King Khalid promised to provide one hundred full-time scholarships to black students to study in Saudi Arabia. The following year, three of King Khalid's well-heeled nephews set up a $50-million investment program for black communities in half a dozen American cities: New York, Washington, Cleveland, San Francisco, Los Angeles, and Atlanta. The investment was to be channeled through the First African Arabian Corporation, a Saudi owned firm on which the three Saudi princes served as directors. Also appointed to the board of directors of the First African Arabian Corporation was Lawrence A. Bailey, the second highest ranking black member of the Carter White House staff.[32] Though the program, called Model Communities, was supposed to fuel economic development and create jobs in black communities, an additional component added a unique religious flavor: Islamic centers and schools were also to be created in which Arabic Studies would be taught.

About the same time the Model Communities investment program was set up, Roy Innis, head of CORE, announced that he had succeeded in getting a commitment for a $40-million Arab investment in Harlem, the large black community in New York City. In the end, neither the $40 million for CORE nor the $50 million for Model Communities ever materialized, though no one knows why.

Another attempt to elicit black political cooperation occurred in early 1979. Six hundred black businessmen attended the First Annual Saudi Arabian—Black American Business Conference at the Century Plaza Hotel in Los Angeles on Saturday afternoon, January 20. Among the handful of featured speakers was Gerald E. Gray, head of the Pan American Steel Corporation. The key to getting a piece of the petrodollar business, he advised the gathering, was to "establish some non-economic relationships that would put us in a better position." Specifically, Gray said, blacks needed to support the Arabs: "When Arabs attempted to boycott companies, we didn't say anything in their support. When Arabs were accused of creating inflation by raising the price of oil, we had a chance to articulate their position . . . we're going to have to be their voice in this country if we expect them to participate in business with us."[33]

But Saudi Arabia had sought a much more direct means of finding someone to be "their voice" in the United States, and their search had succeeded. With the help of oil company officials, the Saudis had obtained the services of one of the most respected and best-connected individuals in the nation's capital.

5

DUTTON
OF
ARABIA

And I want to thank you, too, for
the opportunity to meet Yamani who
I sensed is about as fascinating an
individual as John F. Kennedy, whom
I came to think while his Secretary
of the Cabinet was about the most
interesting political figure (even with
his faults) I would ever come to know.

Frederick G. Dutton,
in a letter to
a high-ranking Aramco
official, December 1973

M r. Dutton loves to trade ideas and he has long been one of Washington's most 'useful reliable' sources for political reporters,'' reported the *New York Times* on April 2, 1978, in a profile of Fred Dutton. The comment was intended not as a compliment, but rather as a simple statement of fact. Irrepressible and filled with energy, Frederick G. Dutton was and continues to be a popular and widely sought after fixture on Washington's political, journalistic, and social circuit. The fact that Dutton registered as a foreign agent of Saudi Arabia in 1975 is viewed by journalists and senators as almost incidental. He carefully maintains a lawyer-client relationship with his clients, the Saudis. And it is that professional distance, combined with his inexhaustible supply of imaginative ideas and his impeccable background, that has sustained his popularity.

In the nine years that Dutton has served as their agent, he has helped build for the Saudis the image of moderation and legitimacy they desperately sought after the ill will engendered by the 1973 oil embargo. The Saudis have paid him handsomely: over $2 million in nine years, including a large expense account. One biannual statement filed at the Department of Justice, for example, shows $400,000 in salary (to Dutton and his wife) and $18,500 in reimbursed American Express expenses, for hotels, meals, and car rental.

Dutton has cultivated friendships with many senior government officials, and with preeminent members of the national media, including *Washington Post* Publisher Katherine Graham, *Washington Post* editor Benjamin Bradlee, *New York Times* Washington Bureau Chief Bill Kovach, columnist Joseph Kraft, NBC-TV anchorman Roger Mudd, CBS-TV diplomatic correspondent Robert Pierpont, and the *Wall Street Journal*'s Karen Elliott House.[1] They, in addition to scores of other media representatives and U.S. government officials, are guests at his frequent brunches and dinners, where the Saudis reap invisible political benefits, such as increased respect and legitimacy, even when they are not the subject of discussion.

Even members of Congress who are not friendly toward Saudi Arabia concede that Dutton has worked miracles for his clients. Said one senator, "When I talk with Fred, I simply never have the feeling that I'm dealing with a foreign agent. I never have the suspicion that Fred Dutton could sell out this country's interest." And, says one prominent journalist, who has attended numerous Dutton dinners, "If Fred Dutton is their man, how bad can they [the Saudis] be?" At the *Washington Post*'s Monday editorial meetings, it's easy to tell who had brunch with Dutton the day before, one staffer said wryly.

According to his statements filed at the Department of Justice, Dutton's work consists primarily of providing legal advice and political counsel. His activities range from giving legal assistance concerning auto accidents and Saudi embassy real estate acquisitions to providing direction to his Saudi clients in utilizing the American political process more effectively. In one twelve-month period, between June 1983 and June 1984, Dutton's registration statements showed extensive traveling on behalf of his Saudi clients: Saudi Arabia (three times), France, West Germany, New York, California, Texas, Massachusetts, Oregon, Georgia, Illinois, Pennsylvania, and Colorado. During that same period, Dutton conferred with Congressman Clement Zablocki (head of the House Foreign Affairs Committee), Congressman William Broomfield, Congressman Thomas Foley, Senator Ted Kennedy, Special Middle East Negotiator Donald Rumsfeld, and Colonel Robert Lilac of the National Security Council. Dutton also sent a detailed, eleven-page letter to every member of Congress in April 1983, in which he asserted that aid to Israel ought to be cut substantially. And he arranged a Washington press conference for the Saudi Ambassador. Indeed, wrote Karen Elliott House in the *Wall Street Journal,* "the heart of Dutton's strategy is his recognition of the central role of the media in shaping foreign policy." [2]

Though he was in private legal practice in 1975 when he signed up with Saudi Arabia, the forty-eight-year-old Dutton had spent much of his life in the public political arena. He was known in Washington for his brilliant insights, acute political suaveness, and unlimited reservoir of new ideas. A graduate of the University of California at Berkeley, Dutton went to law school and became an adviser to Democratic presidential candidate Adlai Stevenson and California Governor Edmund G. Brown. After working in the 1960 campaign of John F. Kennedy, Dutton was appointed special assistant to the President in 1961.

Less than two years later, Dutton moved to the State Department as assistant secretary of state for congressional relations. As one of his friends said, this "proved to be his best preparation for his current role—it gave him hundreds of key contacts." In 1964 he resigned from President Johnson's administration, having been in charge of LBJ's speech-writing as well as the Democratic party platform, and eventually set up his own Washington law firm—Dutton, Zumas and Wise. The associates were all liberal Kennedy family devotees. At one point, Herb Schmertz, now vice-president of Mobil, was an associate of Dutton. When Schmertz moved to Mobil, he hired the irrepressible attorney to provide intelligence and analysis of the "general political climate" in Washington. Other clients included Aerojet General, which wanted to diversify and thus decrease its dependence on government contracts; *Playboy* magazine, whose executives retained Dutton to counter anti-pornography laws; and Norton-Simon, the giant California conglomerate.

As a key adviser to Robert F. Kennedy in 1968, Dutton was at Kennedy's side when he was shot in the kitchen of the Los Angeles Ambassador Hotel. Four years later, George McGovern called on Dutton to assist him in his bid for the presidency.

His commitment to politics was not limited to electoral campaigns. A regent for the University of California from 1962 to 1978, Dutton was a staunch supporter of free speech and also engaged in several battles over policies he felt were racist. One celebrated confrontation occurred in 1970 when Dutton succeeded in angering California Governor Ronald Reagan so much that the governor stood up, pointed a finger at Dutton, and called him a "lying son of a bitch."

Until 1973, Dutton recalled in a 1982 interview with me, he had "never thought about the Arab world, the Middle East, other than a very uncritical pro-Israeli position."[3] In fact, Dutton added, "I had never thought about representing anybody abroad. All my business had been in California."

What caused the change in Dutton's views on Israel? Why did he accept the contract with Saudi Arabia? According to Dutton, it wasn't simply because he received a generous offer. He said that "the original [Saudi] contract was not that lucrative to begin with. It has grown with time, as the amount of work or let's say Saudi trust in me has grown." Dutton only became interested, he said, as a result of Arkansas Senator J. William Fulbright, whom he had gotten to know in the 1964 presidential election: "Fulbright was the one who said that the Saudis needed representation. He and I had never talked about it. It was the most shocking thing in the world when he said that. The Saudis I gathered had been thinking, looking around, and thought of a couple of other lawyers in town who are more prominent, better established than I, but older. . . . The 1973 oil problem had come along, and they suddenly were realizing they had more problems than before. They were a little bit more outward going, and Fulbright at that moment said, what the hell, why don't you have better representation or input in Washington. He was chairman of foreign relations then, and so it got picked up.''

Dutton also added that his decision to take the contract stemmed from a personal connection with Fulbright. ''If almost anybody else had come to me and suggested I do it, no, I don't think I would have done it. It was unusual, but interesting. He was chairman of the Foreign Relations Committee, I admired him, he identified the problem and so I went over there.''

Dutton has long insisted to friends and acquaintances that he was initially hesitant to take the Saudi contract. The *New York Times* gave the following account on April 2, 1978, based on an interview with Dutton:

> *As Mr. Dutton tells the story, the Senator told Mr. Warner [chairman of Mobil] that Saudi Arabians "didn't handle themselves well" in Washington and should hire a representative. Mr. Dutton's name cropped up—among others—and Mr. Warner passed Senator Fulbright's advice on to Riyadh.*

> *In 1975 Mr. Dutton was offered a contract by the Saudis, but hesitated. Given the intensely pro-Israeli position of many in the Democratic Party, he feared that working for the Saudis might seem the end of his political career, but he took the job anyway.*

The real story, however, is a bit different: internal oil company records, correspondence, and new information provided by a former senior Aramco official provide a fascinating glimpse into the way in which Dutton secured his contract and into the nature of some of his public relations activities for his client.

The seeds of Dutton's new career were sown in the middle of 1973 when King Faisal began to lean heavily on the Aramco consortium—Mobil, Exxon, Texaco, and Standard Oil of Cali-

fornia—to initiate political action and lobby in the United States
to change American foreign policy. The Saudis were pleased by
the response from the oil companies, but their appetite for additional political action, having been whetted by the ease with
which the companies caved into initial Saudi requests, grew almost insatiable. It was becoming painfully clear by August 1973
to several officials that their consortium could no longer afford
to act as the chief Saudi lobbyist lest disclosure of this extraordinary—and possibly illegal—role be publicly revealed. Moreover, they needed to deflect some of the Saudi pressure away
from themselves.

Frank Jungers, former head of Aramco, remembered the
problem: "As one of the companies that has the biggest economic interest in the country of Saudi Arabia, we were contin-
ually looked upon to provide the Saudis with advice on a whole
host of matters. I mean from cornflakes to whatever. And I, for
one, and many others, including people like Rawleigh Warner
who were shareholders, felt that this was not the role we should
be playing. And so we were interested in this case and others in
getting them an adviser from the U.S. scene who could advise
them on whatever subject they wanted or could direct their efforts or whatever they wanted in this country."[4]

At the same time, the chairman of the Senate Foreign Relations Committee, Senator Fulbright, told the Saudis and Rawleigh Warner that the Saudis should improve their representation
in Washington by hiring an experienced and politically astute
adviser. The fifty-five-year-old Fulbright had been one of the
leading congressional critics of the Vietnam War, but had also
been an outspoken critic of Israel. (His antagonism toward Israel apparently did not originate with Israel's 1967 occupation
of the West Bank: In 1963, Fulbright organized and chaired a
set of Senate hearings that investigated several American Jewish
organizations as part of a larger investigation into "foreign lobbies.") Fulbright recommended that Dutton, with whom he had
worked on the 1964 Johnson presidential campaign and in sub-

sequent planning against the Vietnam War, be considered for the position of Saudi lobbyist.

In late August 1973, the oil companies approached a half-dozen established Washington attorneys, but none would take the job. In early September, the search narrowed to Dutton, and on September 12, Mobil's Warner—who had taken a leading role in screening candidates—sent Aramco officials a "personal summary" on Dutton "as a possible man for the Saudi Arabians." Soon thereafter, the veteran Kennedy aide met with Mike Ameen of Aramco in Washington. After Ameen explained Aramco's needs, Dutton relayed to Ameen his ideas for a full-scale public relations effort in helping Aramco redirect positive American public opinion away from Israel. According to an Aramco official, Dutton was extremely eager to take the job.

Aramco officials were still in the process of evaluating Dutton when the October 1973 War erupted. The Middle East was now on the front page of every American newspaper and in the top slot on every television news broadcast. Even though the tide of battle turned against the Egyptians and Syrians, they had succeeded in pushing the Arab-Israeli dispute to the head of the world's agenda—especially that of the Americans.

In the early morning hours of October 16, 1973, Frederick Dutton, like so many other members of Washington's political intelligentsia who had been eagerly following the tense drama that was still unfolding in the Middle East, awoke to hear the startling news: Arab oil producers had raised the price of oil 70 percent, and a special delegation of Saudi, Kuwaiti, and other Arab envoys would be dispatched immediately to meet with President Nixon and Secretary of State Kissinger, presumably to demand that Israel be forced back to the 1967 borders. To Dutton, the scent of oil blackmail was disturbing. It was clear to Dutton that this was not the best way for the Arabs to proceed, nor was invoking the oil weapon going to endear the Arabs to the American people. If anything, the Arabs would produce a nasty backlash.

At home, Dutton hammered out a two-page letter to Mike Ameen, a vice-president of Aramco, who was stationed in the oil company consortium's twenty-year-old offices in the Shoreham Building in downtown Washington. Dutton warned Ameen that changes in American foreign policy could not be accomplished by Arab envoys "bearing stern warnings" to the United States. American understanding of the Arab position, Dutton explained, could only be achieved through skillful manipulation of the public policy process.

Acknowledging that he was "still very much on the sidelines" and did not intend to be "pushing myself to help you in this situation," Dutton advised the senior Aramco official that while the delegation of Arab envoys coming to meet with Nixon was a "useful step," their trip "was far from enough as a major move at this juncture—and would seem to misunderstand the overall power structure and real decision-making process in this country, especially now." Dutton continued:

> Genuine effective leverage, at least in the U.S. political system, must not be [through] a major person but through the public, or key groups, or major news sources. The Israelis well understand that. . . . The leverage comes through sympathetic papers, senators, contributors and such. . . . [The Arabs] need to get their case across to key senators, newsmen and others, with them then reaching out more broadly. That will help give Nixon the latitude that he needs in order to move to a balanced policy.

Dutton went on to say that Aramco should initiate the public relations campaign he had previously suggested to Aramco, and solicit help from sympathetic figures in government and the media, such as Senator Fulbright and columnists Evans and Novak:

> It must be backed up with the kind of effort we touched on—and quickly. Fulbright and others should be kept briefed; and so should the newsmen here. . . . How-

ard K. Smith on ABC-TV news last evening (nearly twenty million viewers) made a masterful appeal for a more balanced U.S. policy on the Middle East. The Sunday newspapers all contained columns or articles strongly presenting the Arab view. Yet my inquiries of the people writing those (as Evans and Novak; Smith Hempstone on the Star-News; *and others) is that they are working out of their own background and digging, not with any real or fresh information. . . . In brief, the opportunity is present for the kind of major effort backing up the diplomacy today suggested above.*

Once Dutton knew he was in the final running for the Saudi job, his political conversion on the Middle East seems to have been completed. During the next few weeks, Dutton utilized his press contacts in following up on his suggestions to Aramco. In one memo he wrote to correspondent Murray Marder of the *Washington Post*—as the Israelis were still trying to repel the invading Arab armies—Dutton toned down his advocacy of the Arab position but raised subtle questions about the wisdom of support for Israel. The same man who so courageously had stood up for political and social justice in the 1960s now downplayed the Israeli plight and asked Marder somewhat rhetorically (and misleadingly) whether the American public "would stand for American blood being shed now in the Middle East, or a Middle East tax surcharge being exacted here, or another round of inflation accepted as necessary to confront the Soviets in the Middle East?"

Sometime in late November Aramco agreed to retain Dutton as an adviser. His duties reportedly were to include some of the same services he intended to provide Saudi Arabia: public relations, political intelligence, and help in planning and arranging visits by Saudi officials to the United States and by key American officials to Saudi Arabia. It was both Dutton's and Aramco's understanding that it was only a matter of time before Saudi Arabia would agree to a formal arrangement with Dutton.

On December 7, 1973, Dutton wrote Aramco a rather remarkable letter. It strongly suggests that the genesis of Dutton's pro-Saudi public relations activities in the United States, and his eventual Saudi contract, lay in Aramco's need to prove its loyalty to the Saudis. In his four-page single-spaced letter, Dutton detailed a massive public relations program that Aramco could initiate on its "own," Dutton wrote, "responsive to what I assume is a fundamental need to demonstrate to the Saudis that Aramco is taking real steps here to carry forward your and their common interests." In what would seem like an ironic reversal of roles, Dutton impressed on the oil company executive the urgent self-interest of implementing such a program:

> *The huge stake that Aramco has there would seem to make that especially important over the next three to six months, with the impending negotiations, oil embargo and underlying struggle looming so large. Not making a genuine effort over the months immediately ahead would seem, in fact, to be readily subject to interpretation by them as that the company is ducking in the really critical clinch.*

Dutton's suggestions as to how Aramco could demonstrate its fealty to the Saudis were quite pragmatic—even gauging the likely psychological impact that would result from press clippings. Dutton wrote:

> *I assume that for the purposes of reinforcing relations with the Saudis, such steps as are taken should be susceptible to being readily shown to them. Press clippings are the easiest for that; but the new machines for taping TV shows at home (cost about $1,300) could be used to record what is done here on television; and the tapes then sent for showing there. Sophisticated Saudis like Yamani must be well aware of how much more influential TV has become than*

newspapers with the White House, Congress and
general public; and showing tapes of films can even
make a more indelible impression than thumbing
through batches of newspaper stories.

The common denominator for the series of actions Dutton outlined was that they be "directly verifiable" to the Saudis. One
suggestion was a "special tabloid on Aramco and Saudi Arabia"—Aramco later followed through with this suggestion—for
publication in the *New York Times,* "which could demonstrate
to the Saudis that the company is concerned with getting across
the importance of Saudi Arabia with key elements of U.S. society." Another was a "series of fifteen minute radio 'buys' which
could be taped for English-speaking Saudis to hear."

Another idea was a "unique five-minute documentary on
evening TV news," in which Aramco would "purchase all the
commercial spot-time on the Cronkite Evening News or NBC
for one evening if the network is willing to allow the time to be
consolidated into a single five-minute spot."

And yet another suggestion was that Aramco fund "lectures and seminars on U.S.–Saudi relations" at major universities across the United States, an idea that seems to have been
exploited later by the Saudis and their corporate supporters. In
order to ensure that special forums be held featuring key Saudis
and "select scholars," Dutton urged that Aramco make a grant
to the Middle East Institute for its head, Lucius Battle, "to travel
throughout the nation visiting major newspapers and talking about
Saudi Arabia and the Middle East." Unknown to Dutton, Battle, a former ambassador to Egypt, had independently approached Aramco for funds for his Middle East Institute. In fact,
the institute would soon become a major recipient of Aramco
largesse precisely for its ability to provide the Saudis with a well-
known and credible public relations outlet.

On the last day of November, Sheikh Yamani and his Algerian counterpart, Oil Minister Belaid Abdesselam, commenced a trip to Western Europe and the United States to dis-

cuss the oil embargo. On December 1, the two oil ministers met with Dutch Economics Minister Ruud Lubbers and demanded that the Netherlands—the only country besides the United States to be slapped with the oil embargo—issue a statement in support of the Arabs. But the Dutch government refused to turn against Israel in exchange for getting the embargo lifted.

The next day, the two oil ministers departed for the United States. Arriving in New York, Yamani addressed a press conference on Monday, December 4, and vowed that the embargo would be lifted only when Israel withdrew "from the occupied areas of 1967, including Jerusalem." Meeting with Kissinger in Washington the following day, Yamani reaffirmed the conditions he demanded in return for lifting the oil sanctions.

Aramco officials arranged a twenty-minute meeting between Dutton and Yamani. Dutton also helped Yamani meet the demands of his hectic Washington schedule. Before the oil minister departed, Dutton sent him a gift—a photobook of Washington—at the Madison Hotel where Yamani was staying. In a note accompanying the book, Dutton wrote that it was "a very small token of appreciation for all you have done on this trip, from those who believe in an evenhanded U.S. policy toward the Middle East." He added that he hoped there would be "many more renewing follow-ups on that from many sources."

Two days after he had the book delivered to Yamani, Dutton zipped off a four-page detailed letter to Jungers in Aramco's headquarters in Dhahran, Saudi Arabia. "Yamani's trip here was superb with much good coming of it," Dutton told Jungers, "but much more is needed—and quickly." He outlined a whirlwind coast-to-coast tour, including visits to such places as the Detroit Economic Club and the Los Angeles Town Hall, for a future visit by Prince Saud and Sheikh Yamani.

In his final paragraph, Dutton paid a supreme compliment to Yamani: "And I want to thank you, too, for the opportunity to meet Yamani, who I sensed is about as fascinating an individual as John F. Kennedy, whom I came to think while his Secretary of the Cabinet was about the most interesting political figure (even with his faults) I would ever come to know."

Working feverishly during December, January, and February, Dutton supplied Aramco officials with a steady stream of political memoranda and public relations suggestions. One such suggestion was the publication of a three-page tabloid, complete with pictures and graphics, which was to begin with the salutation "From Saudi Arabia to America" and was to be signed by King Faisal or Yamani. In justifying the oil embargo, Dutton laid blame on the Israeli settlements in the occupied territories and on Jewish immigration to Israel: "The stepped up recent building of permanent settlements on occupied Arab territories and the international campaign launched for still greater immigration to the area inescapably forewarned of yet further attempted expansionism into Arab soil and larger injustices if resolute corrective action was not sought." Dutton held out the promise of Saudi Arabia not only ending the oil embargo but ultimately raising production to 20 million barrels a day as well—if the United States acquiesced to Saudi demands. (Before the embargo, Saudi production was 8 million barrels a day.)

Dutton also conceived a secret public relations "operation" for Saudi Arabia in Washington, which he typed in a three-page outline. Under the aegis of a high Saudi official, the "operation" was to depend heavily on staff personnel from Mobil, SoCal, Texaco, and Exxon. The "staff nucleus" was to consist of a "TV and radio news specialist," "newspaper experts for contact," writers, research assistants, and liaison people with the Senate and House of Representatives. Senators, representatives, and "major news people" were to be invited to Saudi Arabia. Targeted groups included the "business community," "blacks," "environmental groups," "peace groups," "individual VIPs," "governors, mayors, and others."

As the primary contact between Dutton and the Saudis, Frank Jungers would receive Dutton's suggestions and relay them, as appropriate, to other Aramco officials and Saudis. In early January 1974 Jungers thanked Dutton for his three letters and "attachments" and said he had given copies of the first two to Yamani and planned to give them also to Kamal Adham, King Faisal's brother-in-law, whom Jungers described as a "special adviser to the King," though Adham was really the head of Saudi intelligence. Jungers soft-pedaled Dutton's suggestion for a "brainstorming session" with Yamani as "premature," but said that he would show Adham his suggestion and "press him [Adham] to consider additional action."

Dutton insisted on being exclusively in charge of the operation and working directly with Kamal Adham and top Aramco officials. Quoting a fee of $1,000 a day for top Washington lawyers, Dutton asked for a three-year contract for $600,000 and added, "I will likely be giving thirty days a month for a long time to come, and long evenings." The contract would be subject to cancellation "if Fulbright informally indicates [that I am] not performing or for other grounds." (This was a rather interesting role for the chairman of the Senate Foreign Relations Committee to play. No evidence exists, however, that Fulbright actually reported on Dutton's activities to the Saudis.) In the final line of the outline, Dutton wrote, *"Start Immediately*—should have started six months ago, or twenty-five years ago" (the year Israel was created).

In January and February 1974 Dutton met with Adham's nephew, Prince Turki—who has since succeeded Adham as head of Saudi intelligence—several times, in the United States and in Saudi Arabia. He was, however, unable to meet Adham, and no contract was signed. At the conclusion of Dutton's two-week visit to the kingdom in early February, Dutton wrote Jungers expressing his appreciation for making the trip possible. He hoped that "a working relationship with Kamal Adham goes together as it provides striking opportunities for getting at the real problems of these times both here and more broadly."

By February, Dutton had been led to believe that the Saudis would finally agree to a formal contract. So he drew up a "Statement of Understanding" with eight clauses enumerating his duties as a "consultant," rates of compensation, method of payment, and expense reimbursement. The contract was to begin on February 15, 1974. Dutton stipulated in clause two that he would "be guided by and report to Sheikh Adham and shall not be assigned to work regularly under a ministry or bureaucracy, nor under other Americans.'' The contract was to run for three years ''because of the almost certainly protracted nature of the problem to be dealt with concerning Saudi Arabian–U.S. ties and because other sources of income and work for me will be affected by my taking this responsibility.''

In another memo, Dutton dwelled extensively on the importance of his close friend Senator Fulbright as ''the only major voice in the U.S. trying to educate other political figures and the general public as to why American policy should be even-handed in the Middle East.'' In fact, Dutton wrote, Fulbright and Secretary of State Henry Kissinger, despite their antithetical views on the Vietnam War, formed an unusual alliance on the Middle East. The Arkansas senator ''privately provided much of the backbone in Kissinger's own thinking about the region.''

Moreover, Kissinger engaged in some rather extraordinary steps to help Fulbright in his troubled bid for reelection. The young, popular governor of Arkansas, Democrat Dale Bumpers, had been dropping rather strong hints that he intended to challenge Fulbright in the 1976 primary. Kissinger, Dutton revealed, had ''flown to see him [Fulbright] there on February 16th [an almost unheard of courting by a U.S. secretary of state] in order to get Fulbright's thinking before going on to see Nixon in Florida the same evening to develop a U.S. posture toward Ommar Saqqaf [the Saudi foreign minister]. Kissinger made explicit just by his quick trip to Arkansas that he needed to sound out Fulbright before Nixon and the State Department could map out their position.'' Dutton continued to provide up-to-the-minute status reports on Fulbright—who was greatly admired by Aramco officials—attributing the Arkansas senator's reelection difficulties to the support given to Fulbright's opponent by ''pro Israel groups.''

In the absence of a contract with the Saudis, Dutton was retained by Aramco. In a memo sent to Jungers in late February, Dutton supplied background material on hearings—scheduled for the spring—into the practices of the oil companies by the Senate Subcommittee on Multinationals of the Senate Foreign Relations Committee, chaired by Frank Church.

Dutton also related a "small vignette" that indicated to him that "there is some dark thinking going on in Washington." On the evening of Friday, February 22, 1974, Dutton was scheduled to have a "quick informal dinner" with CBS reporter Fred Graham and David Halberstam, a Pulitzer Prize winner for his reporting on the Vietnam War and author of *The Best and the Brightest*. The three of them dropped by a hotel to see Eric Sevareid, the CBS commentator, and Theodore White, author of the much acclaimed series of books, *The Making of the President*. Dutton mentioned to White, whom he had known for a long time and had grown close to during Robert Kennedy's presidential bid in 1968, that he had recently visited Saudi Arabia. White, in Dutton's words, "immediately assailed me and began a long monologue on how no little country of 2 million people (he even had his figures wrong) should be able to cause so much trouble to all the industrial nations in the world. And Israeli paratroopers should be sent in!" White continued in a

> warped, distorted, incredible diatribe by what is supposed to be a highly intelligent man and certainly an influential one, including more with U.S. political leaders (especially Nixon) than with the reading public. No argument or moderating could change his mind. The scene was frankly grotesque. But it is mentioned here because I have run into varying degrees of it with respected political writers at several Washington dinners and several of the younger editorial writers at the Washington Post.

While awaiting a formal agreement, Dutton provided services to the Saudis in 1974 for which two high-ranking Aramco officials thought Dutton should be paid. They urged that he submit a bill to the Saudis in May. Even though this arrangement was apparently agreed to by Adham, Ameen advised against it lest the

Saudis "misinterpret" the bill. "I feel it best," Ameen wrote Dutton, "that we continue to attempt to have him [Adham] enter into an agreement signed, sealed and delivered."

Dutton was getting paid for his work—between March 1 and June 1, 1974, he billed Aramco for 20,200 dollars' worth of work—and he was earning a solid reputation among the Saudi elite. For example, the Saudi billionaire and businessman, Suliman S. Olayan, retained Dutton in his effort to get appointed to the board of directors of American Express. In recognition of the vast new financial power of the Saudis, American Express had been looking to appoint a prominent Saudi to its board. In a letter to Dutton on May 10, 1974, Olayan rejected Dutton's suggestion that he ask Rawleigh Warner of Mobil to intercede with American Express in his behalf. Olayan preferred to wait and see what transpired from his own low-key approach to Howard Clark, a director of Mobil, when Olayan accompanied Clark to Saudi Arabia in late April.

Although Dutton was increasingly popular with other Saudis as well as with the entire Aramco hierarchy, his contract with the Saudis still proved elusive. On May 27, Mike Ameen promised Dutton he would finally confront Adham, who was going to the United States in early June, to "get this matter settled once and for all." Once again, contact proved impossible. Dutton was invited back to Saudi Arabia in July, where he finally met with Adham and with Prince Turki. Saudi deliberations about hiring Dutton proceeded slowly—a delay that Jungers attributed to "normal Saudi procrastination"—but by the end of 1974, they finally decided to retain him for matters concerning oil. A formal contract was negotiated, and on April 7, 1975, Dutton deposited in his bank a $100,000 fee from Petromin, the Saudi state oil agency, also known as the General Petroleum and Mineral Organization for Saudi Arabia.

On June 10, 1975, Dutton registered at the Foreign Agents Unit of the Department of Justice—nearly a year and a half after he secretly began working for Saudi Arabia—as an agent for Petromin. Sheikh Yamani was designated principal contact for

Dutton. Six months later, Dutton again registered at the Department of Justice, this time as an agent for the embassy of Saudi Arabia in Washington. In the same way he has assiduously maintained his reputation as a man who is above crass influence peddling and manipulative public relations, Dutton stipulated that he was "being retained for counseling and legal advice, not public information activities, which registrant understands will be the responsibility of others." Dutton added that he intended "to protect and assert his right as a U.S. citizen to speak out concerning his convictions in the same various ways he has in the past before undertaking this counseling." Dutton would later establish himself as one of the preeminent foreign agents in the United States.

6

THE BEST
AND
THE BRIGHTEST
THAT
MONEY COULD
BUY

I doubt we would accept [Arab] boycott advertising. Still, I guess it is
hard when your client is a king. What
do you say when the copy is lousy—
''Gee, King, uh . . .''

Advertising executive
talking about his firm's
PR plan for OPEC

As a result of Dutton's contract with the Saudis, the floodgates burst open to Washington officialdom and public relations experts seeking to sell their contacts, prestige, and expertise. Within the first three years after Dutton signed up, he was joined by scores of illustrious former high-ranking government officials offering priceless prestige and legitimacy and an invaluable political platform to a handful of previously scorned nations. Petrodollars were turning out to be a commodity that could purchase not just washing machines and automobiles, but also prestige and favorable public opinion. The modern-day gold rush had begun.

One massive public relations program, for example, was designed right after the October 1973 War. After six months of prodigious research, a 20,000-word confidential document was finally completed on March 31, 1974. In extraordinary detail, the ''Public Affairs Program for the Arab World'' unveiled a meticulous blueprint for manipulating American public opinion, the press, and the political process. Among the plan's numerous components was one section that laid out a strategy for defeating six senators—by channeling contributions to their opponents, providing advertising for them, mobilizing local voters, and doing other grass-roots campaigning. Those six senators were considered "adversaries" of the Arab cause: Indiana Democrat Birch Bayh, Florida Republican Edward J. Gurney, Ohio Democrat Howard Metzenbaum, South Dakota Democrat George McGovern, North Dakota Republican Milton Young, and Kansas Republican Robert Dole.

The propaganda plan, later disclosed by national columnist Robert Walters in *Parade* magazine, was produced by Martin Ryan Haley and Associates, Inc., a well-established New York–based public relations firm with offices in Washington, Rome, and Brussels.[1] The company was headed by Martin Ryan Haley, a political consultant who had worked on more than eighty national and state political campaigns.

Another section of the plan called for massive political advertising in thirty major metropolitan newspapers (to reach 140 million readers), seven hundred local television stations, and two to four thousand radio stations. Martin Ryan Haley wanted $15 million annually for this three- to five-year program, which would utilize over a hundred staff members. Haley did not find a buyer for his program and consequently it was never implemented.

Martin Ryan Haley was just the first of many professional public relations specialists to offer their services to oil producers. Unlike Dutton, these career public relations specialists could not offer immediate prestige or access to decision makers. But like Dutton, they successfully played on the Arab nations' unfamiliarity with, and determination to change, American policy.

Arab leaders were repeatedly told that changes in American public opinion were effected by the behind-the-scenes manipulations of influence peddlers and by propaganda schemes. Arab leaders could only be led to one inescapable conclusion: altering American policy and popular attitudes was simply a matter of pulling the right strings.

In December 1974, another widely respected public relations firm had made an effort to implement a public relations program on behalf of OPEC. The well-known New York firm,

PKL Advertising, Inc.—which had done some much-admired work for the late Robert F. Kennedy—offered OPEC a $10-million "campaign of information" in television, newspapers, news weeklies, and radio. The purpose? To counteract the "smear job" performed by the American media and to "spell out the goals of the OPEC countries, tell of their conditions following the price of oil, the possible offering of scholarships to colleges for study in oil technology, history of Arab nations, etc." Guaranteeing that the information would reach 95 percent of "adult Americans," PKL said its program would "lessen the possibility" that the growing American backlash against OPEC would develop into any "decisive political, technological, or economic action." OPEC officials then requested approval for a six-month contract.[2]

The OPEC/Arab accounts were obviously considered quite valuable. Within the next month, according to published reports, another approach was made—this time by *Reader's Digest.* Three top representatives of the magazine met for several hours with OPEC Secretary General M. O. Feyide, detailing their plan for a mammoth public relations campaign in the United States to "correct the wrong, one-sided, and bad image which has been given to OPEC."[3]

The highly unusual and very discreet meeting took place in Vienna on January 17, 1975, in the plush OPEC headquarters in the Austrian capital. Guaranteeing to reach "millions of middle class type of intelligent and thoughtful readers," *Reader's Digest* officials offered to publish a dozen favorable articles in their magazine over the course of a year for a fee of up to $4.53 million. OPEC's secretary was so impressed with the idea and presentation that within eight days he circulated a memo to OPEC ministers asking them to approve a "limited" contract for four commissioned "articles."

When asked about these proposals, which were revealed in *Business Week* and *Platt's Oilgram News,* spokesmen for both PKL and *Reader's Digest* tried, rather unconvincingly, to put the best light on the obviously embarrassing disclosures. PKL's

President John Shimma claimed, "I would doubt we would accept [Arab] boycott advertising." Then he added, "Still, I guess it is hard [to turn advertising down] when your client is a king. What do you say when the copy is lousy—'Gee, King, uh . . .' "[4]

The managing editor of *Reader's Digest*, Edward T. Thompson, claimed that the magazine was only offering advertising. "They want to buy pages. We'll sell them pages." But as *Business Week* pointed out, one of the officials who traveled to Vienna was a European editor with no responsibility in advertising.

By the end of 1975 there was no shortage of former American officials and public relations experts available to do the bidding of Arab governments in the United States.

Defeated in his bid for reelection in 1974, J. William Fulbright, the man who originally helped Dutton land his contract with the Saudis, had his own reasons to be thankful for his close connections to Saudi Arabia and the oil companies. His stellar thirty-two-year congressional career, during which time he had opposed the Vietnam War, had come to an abrupt end. Though out of office, he maintained a busy public speaking schedule. In an address before the Middle East Institute on October 3, 1975, in Washington, Fulbright blasted the "extraordinary power" of the "Israeli lobby" and called for redirecting American policy toward the Arabs, where American "strategic" interests lie.[5]

In early 1976, William Fulbright, the former Senate Foreign Relations Committee chairman, embarked on a new, prosperous career. He and his prestigious Washington law firm of Hogan and Hartson were retained by the United Arab Emirates (UAE) for $25,000 a year; in mid-1976 Saudi Arabia signed on Fulbright for an additional $50,000 a year plus expenses.

Fulbright had been made personally aware of how enticing Arab wealth could be. In 1972 Fulbright's wife, Betty, received emerald and diamond jewelry worth thousands of dollars from the Abu Dhabi oil minister. Contrary to the law—which had been initiated by the Foreign Relations Committee—requiring government officials to immediately hand over gifts from foreign heads of state, the jewels were not given to the State Department until a year and a half later.[6]

Among his numerous activities on behalf of his Middle East clients, Fulbright hosted dinners for visiting Arab dignitaries and U.S. senators, such as the one held on July 19, 1976, for the Saudi minister of foreign affairs, Senators Herman Talmadge and Patrick Leahy, and Congressmen Robert Krueger and David Obey. He also attended business conferences with leading corporate officials regarding trade with Saudi Arabia, and he produced legal memoranda on the Treasury Department's anti-boycott guidelines. Though his access to Washington's elite was surely valued by Saudi and UAE officials, of even greater value was the political platform that Fulbright provided. The former senator had risen to such eminence that he was considered beyond suspicion—and this of course ensured the legitimacy of his political views. After all, would a retired, distinguished member of the Senate, who had chaired one of its most prestigious committees from 1959 to 1974, sell his integrity for a price?

The press, however, by succumbing to the notion that such a man was above suspicion, overlooked the question of Fulbright's objectivity. One flagrant example occurred in September 1978, immediately preceding the historic Camp David negotiations involving President Jimmy Carter, Egyptian President Anwar el-Sadat, and Prime Minister Menachem Begin.

Fulbright was one of "five experienced observers of Middle Eastern affairs" asked by *Newsweek* to provide their "assessments" of the "chances for progress toward peace" at the soon-to-be-held Summit.[7] In his column for *Newsweek* international section, Fulbright called for a Middle East settlement to be "guaranteed" by the world's superpowers, which would "enforce the peace and all of its specifications." Israel has long distrusted such "international guarantees" and has insisted on face-to-face negotiations leading to security and normalized relations with its Arab neighbors. Arab governments, however, have viewed superpower guarantees favorably because such guarantees (which are tenuous, at best) can wrest concessions from Israel without requiring the Arabs to make their own concessions regarding Israeli security. Fulbright wrote that "guarantees" would be provided by the United Nations Security Council— hardly a neutral or even effective force, considering its long history of consistently adopting one-sided anti-Israeli resolutions— with "responsibility for enforcement" carried out by the five permanent members: the United States, the Soviet Union, France, the United Kingdom, and China. The Soviet Union and China supply substantial military equipment to Israel's adversaries and do not even recognize Israel, Fulbright neglected to point out, nor did he mention France's and Great Britain's unmistakable pro-Arab tilt.

In its byline, *Newsweek* identified Fulbright as a "former U.S. Senator [who] practices law in Washington, D.C." Nowhere did *Newsweek* mention Fulbright's status as a foreign agent of Saudi Arabia. Nor—and this is even more important—were readers made aware that Fulbright's article was in fact pure, certifiable propaganda on behalf of Saudi Arabia. On his mandatory biannual report of his status as a foreign agent for Saudi Arabia, which Fulbright filed at the Department of Justice in February 1979, he responded yes to this question: "Have you ever on behalf of a foreign principal engaged in political activity?" Fulbright then listed the *Newsweek* article, an article in the now defunct *Washington Star,* and a speech on the Middle East before the Council on Foreign Relations in New York City.[8]

Though having Dutton and Fulbright on the Saudi payroll surely represented quantum leaps in advancing that country's public relations in the nation's capital, two major additions were made in early 1978 that broadened the kingdom's access to the Democratic and Republican parties and the business community. J. Crawford Cook, the head of a South Carolina–based public relations and political consulting firm—Cook, Reuf, Span, & Weiser—was awarded during the last week of March two hefty Saudi contracts: one for $65,000 to lobby for President Carter's proposed sale of F-15 fighter planes to Saudi Arabia, and another for $100,000 as a "down payment" to implement a "long range cultural, educational, economic, civic, and social" public relations program in the United States to enhance the Saudi image. Cook was awarded the contract as a result of the entrée provided by his good friend, the American ambassador to Saudi Arabia, John C. West, beating out several top New York firms also vying for the lucrative contract.

A widely respected political consultant, the forty-five-year-old Cook had previously worked as administrative assistant for Senator Fritz Hollings and had exhibited considerable talent while working in the campaigns of Democratic senators Walter Huddleston and Wendell Ford of Kentucky and Robert Morgan of North Carolina, and Virginia Lieutenant Governor Chuck Robb. For the F-15 contract Cook and his firm diligently courted members of Congress, key staff members, and the press, sending out slick packets of "information," and also contacting over one hundred corporate chieftains in the days prior to the congressional vote. Despite overwhelming opposition at the outset of the debate in March 1978, the sale was approved resoundingly two months later. Up-to-the-minute political intelligence on the Senate deliberations, according to sources connected to Saudi Arabia, was also passed along discreetly to Cook by a staff member of the Senate Foreign Relations Committee.

The sale was approved by a 54–44 vote in May. But the Saudis were undoubtedly also impressed with the fact that both South Carolina senators, Democrat Ernest Hollings and Republican Strom Thurmond—who had previously been considered staunchly pro-Israel—voted for the sale of the weapons. Thurmond's support for the sale was especially remarkable. He had been set to vote against it because he felt the sale would be "destabilizing to the peace efforts in the Middle East."[9] He had even planned to announce his opposition to the sale of the F-15s at a midmorning news conference on the day of the vote. Suddenly, and without warning, the press conference was canceled. Hours later, Thurmond voted for the sale of the F-15s, and Cook was widely credited with having produced Thurmond's change of heart.[10] Within the next four years, these and other behind-the-scenes efforts by Cook were increasingly appreciated by the Saudis. Cook's payments were increased to $470,000 annually.

Cook later played pivotal roles in mobilizing broad-based corporate support for the sale of the super-sophisticated AWACs surveillance planes, monitoring political developments on Capitol Hill, and regularly providing briefings to television and print journalists on behalf of his client. Cook also delivered to Saudi Arabia three separate, extensive public relations programs for implementation in the United States, and produced a slick, expensive Saudi film for exhibition at the 1982 Knoxville World's Fair and elsewhere throughout the United States. Another of Cook's duties, according to his Justice Department registration form, is to "broaden the participation of the Kingdom in Arab-American organizations."

Cook has become a believer in Saudi Arabia, strongly defending its foreign policy and telling me in an interview that "the Saudis want peace more than anything else in the world." He speaks of Saudi modernization in glowing terms: "What has occurred in Saudi Arabia in fifty years may very well be the most remarkable sociological achievement in the history of known civilization." Cook believes that the Saudis are "buried under the Arab-Israeli conflict and OPEC." His aim is to "humanize Saudi Arabia" by exposing the American public to the Saudi achievements, "which are inuring to the benefit of the whole world," such as their desalination projects. When asked about the Saudi's use of the oil weapon, Cook said, "Many other countries use whatever resources are available to accomplish a certain goal. Why should they [the Saudis] be any different?"[11]

At about the same time they hired Cook, the Saudis also retained Stephen N. Conner, a former vice-president of Merrill Lynch Pierce Fenner & Smith in charge of international merchant banking. After leaving Merrill Lynch, Conner became a partner in Manara Ltd., a privately held business consulting firm designed to promote trade and investment between Saudi Arabia and the West. Manara had been formed by Raymond H. Close, a seven-year CIA station chief in Jidda, Saudi Arabia, upon his resignation in February 1977.[12] The thirty-eight-year-old Conner, who had strong ties to both the Republican party and the corporate world, was awarded a $50,000 annual retainer. He immediately began lobbying various senators, congressmen, and their staffs, as well as members of the business community on behalf of the sale of F-15s to Saudi Arabia. Since then, other activities have included meeting extensively with Reagan Administration officials in matters of "mutual interest between Saudi Arabia and the United States," including the Israeli invasion of Lebanon, the proposed U.S.-Israel Free Trade Area, and military sales to the Saudi kingdom. On one occasion, in 1983, according to the Department of Justice foreign agent files, Conner provided questions to senators Strom Thurmond and James Exon (Democrat of Nebraska) to be submitted to officials of the Reagan Administration about a former staff member of the Senate Foreign Relations Committee accused of leaking classified material to the Israeli government.[13]

One confidential public relations document—apparently prepared by one of the American advisers to Saudi Arabia, though the true author has never been established—surfaced in early 1978. The seven-page single-spaced document warned nefariously of the "Eastern Jewish establishment" and the "pro-Israeli forces" acting to "undermine President Carter and his cabinet to make him a detriment in the coming elections and laying the groundwork for his defeat in three years."[14]

In order to mobilize grass-roots support for Saudi Arabia in the United States, the author of the document advocated Saudi visits to various American communities to demonstrate Saudi Arabia's economic importance, and more significantly, urged Saudi Arabia to announce an "all-Arab peace proposal." The purpose of this peace plan was not to break the diplomatic deadlock, but to "outline in sharper detail the hard-core intransigence of Israel," as the author of the document pointed out. Whether this plan served as a partial motivating factor is not known, but in 1982 Saudi King Fahd announced an eight-point "peace" plan. A distillation of previous Arab statements and also U.N. resolutions, the plan did not explicitly acknowledge recognition of Israel's right to exist; but it did call for Israel's immediate evacuation from all the occupied territories and the creation of an independent Palestinian state. Israel immediately rejected the plan. As correctly anticipated, the "peace plan" did not move the negotiations stalemate off dead center, but did highlight Israeli intractability.

Even with such well-connected and politically balanced operatives as Dutton, Fulbright, Cook, and Conner, Saudi Arabia continued to look for additional connections, especially those with ties to whoever was in power in Washington at that particular moment. Less than a week after dark-horse presidential candidate Jimmy Carter emerged as the winner in the February 1976 New Hampshire primary, Saudi Arabia hired Carter's polling firm, Cambridge Reports, headed by Patrick Caddell. For $50,000 Caddell was obligated to give the Saudis "an oral presentation of data"; and for an extra $30,000, the Saudis were allowed to insert thirty questions into Caddell's 300-page, 350-question survey given to two thousand carefully and scientifically selected Americans. The fees that Saudi Arabia paid—it was the only foreign government to subscribe to Caddell's quarterly Cambridge Reports—were twice as much as what Caddell was receiving from other clients, including Westinghouse, Arco, Amoco, Sun Oil, Shell, Ford Motor, and Aetna Life Insurance.[15]

Fifteen months later the Saudis hired Michael W. Moynihan, who operated a Washington public relations firm and who just happened to be the brother of New York Democrat Senator Daniel Patrick Moynihan. Michael Moynihan was retained for one month for a fee of $15,000 to provide background material to the media in connection with the May 24 visit of Crown Prince Fahd to the White House.

The obsession with Israel even led Saudi Arabia to subsidize the work of an American Nazi, William N. Grimstad, who had previously worked as managing editor of *White Power,* the party newspaper of the National Socialist White People's party. The Saudi embassy in Washington, which had maintained a close relationship with him over several years, gave the forty-year-old Grimstad $20,000 as a "gift." According to Grimstad's registration at the Department of Justice, the money was "in appreciation of publication of my 1976 book, *Antizion*," a rabidly anti-Semitic tract. On behalf of Saudi Arabia, Grimstad's activities were to consist of "exposing Zionist imperialism [which would] benefit not only foreign nations, but all humanity." Grimstad notified the embassy that he intended to use the money to research "the illegal usurpation of Palestine" and to incorporate the material in either a "screenplay or teleplay, to be produced for exhibition primarily to students," or a "book."

7

THE PETROCORPORATE CLASS

I see no reason why nearly half of
the foreign aid this nation has to
give has to go to Israel, except for
the influence of this Zionist lobby.

Former Vice-President
Spiro Agnew

Lobbyists and public relations consultants were not the only advisers attracted to Saudi Arabia and other Arab oil-producing countries. Arab oil money soon became a magnet for individuals whose only interest was to enrich themselves by becoming part of the petrodollar recycling process. Throughout American business circles, rags-to-riches stories abounded. Moreover, the flow of Arab investments and level of trade had risen to such staggeringly high levels in so short a period of time that the Arab nations found themselves in urgent need of financial advisers to manage their money. As a result, a new American class of consultants and entrepreneurs—the petrocorporate class—began to materialize.

Throughout American history, new entrepreneurial classes have constantly emerged, usually taking advantage of historical, political, or technological developments for which they were not responsible: the traveling salesmen, the middlemen who provided consumer goods to the newly populated West, or the traders who gained access to China in the late 1880s. The petrocorporate class was generally no different: in providing technical and financial advice, bringing key business contacts together, arranging business deals, and pairing off investors, the petro-corporate class fulfilled a classic "middleman" function.

Yet, two major differences began to set the petrocorporate class apart from all previous entrepreneurial classes: the petro-dollar was not just a financial currency; it contained a political value as well. And while many petrocorporate consultants may have tried to keep the two currencies separate, several of them flagrantly exploited the political dimension. The second major difference concerned the identity of the petrocorporate class. For the most part, Arab nations selected former high-ranking gov-ernment officials who offered not only technical expertise but political access and prestige as well.

And so, through an accident of fate, those government of-ficials who had been the first to know Arab government and fi-nancial leaders—usually in a favorable way—through their gov-ernment service were initially selected as advisers.

Gerald Parsky, assistant secretary of the treasury for mon-etary affairs in the Nixon-Ford administration, traveled exten-sively to Saudi Arabia, Kuwait, the United Arab Emirates, and Qatar in 1974, 1975, and 1976. Conferring with his governmen-tal counterparts in these countries, as well as the managers of their economic and banking institutions, Parsky's main task was to ensure that petrodollars would be directed toward the United States. Only thirty-three years old in 1975, Parsky became known as the whiz kid in the Treasury Department, able to recall and recite country-by-country revenue data, Saudi development plans, Kuwaiti investment figures, and extrapolations of future oil rev-enue earnings.

He also negotiated various economic agreements with Arab governments and played a key role in setting up the Joint Eco-nomic Commission with Saudi Arabia. According to Kuwaiti officials, Parsky apparently assured Kuwaiti government leaders that the United States would never tax Kuwaiti real estate in-vestments in the United States. After he left office, however, no records could be found to confirm the substance of oral assur-ances to the Kuwaitis. This proved to be problematic when the

Internal Revenue Service proposed in 1978 to change a section of the tax law that had previously exempted foreign investment in U.S. real estate. Kuwait protested the change, threatening to reduce oil production.[1]

Not unexpectedly, Parsky emerged as one of the most enthusiastic supporters of Arab investment in the United States, making his case often before American business groups. But Parsky also became one of the Ford administration's most ardent opponents of anti-boycott legislation, frequently echoing the arguments that Arab officials had made in testimony before Congress and in interviews with the media.

In an interview with *US News & World Report* in November 1976, for example, Parsky defended the boycott, saying it was the Arabs' "sovereign right to employ a non-violent tool that they believe has been used by many countries in a state of hostility including the United States." Parsky neglected to mention that the secondary and tertiary components of the boycott rendered it unique and separate from all other types of international trade sanctions.[2] Arguing against any legislation designed to combat the boycott, Parsky said, "the only way to end the Arab Boycott is to . . . resolve the [Middle East] conflict." He acknowledged, however, that the solution could take "years." And what would happen if Congress passed anti-boycott legislation? he was asked. Not only would trade relations with the Arab world be "damaged," Parsky advised, "but antiboycott legislation [will] seriously hurt our efforts to achieve peace in the Middle East."

Thirteen months after the interview was published, Saudi Arabia and the United Arab Emirates showed their appreciation of Parsky: he and his law firm of Gibson Dunn & Crutcher— Parsky became head of the firm's Washington branch after Ford's 1976 defeat—were hired as registered foreign agents for Saudi Arabia and the United Arab Emirates. Parsky's duties included studying American tax policies and lobbying in support of changing tax laws to be more favorable toward his clients. In the next two and a half years these accounts generated $350,000 in revenues to the law firm, which is among the largest in Los Angeles, its home base.

In the year following Reagan's election in November 1980, Parsky's firm—whose partner William French Smith was tapped for the position of attorney general—opened an office in Riyadh. Parsky's current clients include the University of Riyadh and the Saudi Public Transportation Company, which operates a rapid transit bus system.

A contemporary of Parsky, Willis C. Armstrong, assistant secretary of state for economic affairs, also had the good fortune to become friendly with major Arab investors. After his departure from government, he was hired by Saudi businessman Adnan Khashoggi to monitor political and economic developments in the nation's capital. And an even more prominent official from the Nixon administration, William E. Simon, who served as secretary of the treasury from 1974 to 1977 and was Parsky's boss, went to work as a middleman for Arab investors.

Widely respected as a financial wizard before entering government, Simon had spent twenty years on Wall Street, the last few at the helm of the government bond department at Salomon Brothers. As head of the Treasury Department during the oil price "revolution," Simon personally helped negotiate various financial agreements with Saudi Arabia.

One such agreement, the origins and specific terms of which were not revealed until four years after it was secretly negotiated, was an astounding concession to Arab regimes: it required the United States to keep secret the magnitude of Saudi investments in the United States in exchange for an agreement by the Saudis to purchase a large number of government securities, such as Treasury bills. Ultimately, as a House subcommittee would discover, this 1975 oral agreement proved to be one of the most controversial agreements in the area of foreign investment. Simon also opposed any anti-boycott legislation, testifying before Congress that "peace in the Middle East is the only ultimate answer," and that "heavy-handed measures which could result in direct confrontation will not wash."

In 1979, the fifty-two-year-old Simon, already on the board of directors of such leading companies as United Technologies and Citibank, was hired as a consultant to Crescent Diversified, Ltd., the American investment company of Saudi financier Suliman S. Olayan. Within a year, Simon was elevated to chairman of Crescent Diversified and was helping to manage portions of Olayan's multibillion-dollar investment portfolio.

Saudi Arabia was not alone among the Arab oil exporters in securing the services of well-connected personalities. Richard G. Kleindienst, attorney general during the Nixon administration, signed up with another Arab oil-producing country in late 1973. Earlier in that year, Kleindienst had been convicted of perjury for his role in the ITT campaign finance scandal.

Back in private practice after suffering the ignominy of a one-month suspension from the bar, Kleindienst began searching for clients. Ironically, this dyed-in-the-wool, conservative, pin-striped Republican went to work for Algeria, a radical and self-styled "socialist" regime that had broken diplomatic relations with the United States in 1967 and was considered implacably anti-American. Algeria had even occasionally given sanctuary—sometimes providing heroes' welcomes—to Palestinian guerrillas. In April 1973, the United States had even lodged a formal complaint with Algeria for allowing the Voice of Palestine radio to broadcast exhortations to "kill and assassinate anyone who is American."

On September 16, 1973, the fifty-year-old Kleindienst flew to Algiers, at the Algerians' expense, where he met with various Algerian officials including its oil minister, Belaid Abdessalam. The week before Kleindienst's arrival, Algerian President Houari Boumediène, had railed against "Western imperialism" and multinational corporations in a speech before a conference of Third World leaders meeting in Algiers. But that did not stop Abdessalam and Kleindienst from reaching an agreement on September 24 for a $10,000-a-month contract. Nor did the cataclysmic events of the October War or the Arab oil embargo stand in their way. In fact, Algeria was among the most vociferous of Arab states in voicing anti-American sentiments. And when Saudi Arabia was about to end the embargo in March 1974, Algerian President Boumediène continued to demand the maintenance of oil sanctions against the United States.

On the first day of November 1973, Kleindienst formally commenced work for the Algerian out of his K Street office in downtown Washington. His duties included providing legal representation to Algeria and to Sonatrach, the Algerian state petroleum company. Kleindienst also served as an intermediary in setting up meetings between officials of Algeria and American business leaders and executive branch officials such as Secretary of State Kissinger, Treasury Secretary George Shultz, Deputy Secretary of the Treasury William Simon, and then-Congressman Gerald Ford. He also occasionally tried to engender goodwill in Congress for Algeria, by arranging receptions for scores of senators and congressmen on behalf of Abdessalam. One effort involved having the Algerians pay for a private luncheon in September 1975 on Capitol Hill hosted by Republican Congressman John J. Rhodes, the House minority leader, on behalf of visiting Algerian leaders, including its oil minister. An invitation to visit Algeria was extended to Rhodes; the congressman tentatively accepted, but deferred going. On another occasion Kleindienst arranged a reception on behalf of Abdessalam, which twenty-four senators attended.

The consulting firm of Henry Kearns, former chairman of the Export-Import Bank—a government agency that subsidizes American exporters through long-term low-interest loans to foreign purchasers—signed up with Algeria's Sonatrach in 1975, exactly one year after he left government service. In his first year alone, Kearns's firm was paid $350,000.

An even more remarkable individual to sign up with the Algerians was the venerable Clark Clifford, the elder statesman of the Democratic party who had served in the administrations of Truman and Kennedy, and had been Johnson's secretary of defense. Even more than Kleindienst's contract, Clifford's alignment with the Algerians was highly extraordinary because of his unusual background. In 1948, as special counsel to President Truman, Clifford played an instrumental role in persuading the United States to recognize the newly created state of Israel. In a memo to Truman, Clifford had scoffed at threats issued by Saudi Arabia and other Arab leaders: "The fact of the matter is that the Arabs must have oil royalties or go broke. . . . Their need of the United States is greater than our need of them."[3]

But in 1970 the Algerians offered Clifford's law firm an annual retainer of $150,000. In the next seven years, Clifford's firm would collect over $1 million from Algeria for services and expenses. The firm's duties ranged from negotiating loans on behalf of Sonatrach to advising the Algerian government on foreign issues. At no time, however, did Clifford ever pander to Arab sentiment in the Arab-Israeli dispute.

Acknowledged to be one of Washington's most experienced, successful, and expensive lawyers—charging clients $350 an hour—Clifford soon made other lucrative arrangements with powerful Arab interests. He was named chairman of Financial General Bankshares, Inc., a $2.2-billion bank holding company that owns 12 banks—and 150 branches, many of them First American banks—in New York, Maryland, Tennessee, Virginia, and the District of Columbia, after Financial General Bankshares was taken over by Arab investors in a protracted ten-year legal battle. These new investors were Kamal Adham, the former head of Saudi intelligence; Faisal al Fulaig, the former chairman of Kuwait's national airline; and Crown Prince Mohammed, the son of the president of the United Arab Emirates.

Other directors appointed to the board of the holding company included Stuart Symington, former Democratic senator from Missouri and secretary of the air force; and Elwood R. Quesada, former head of the Federal Aviation Administration and special assistant to President Dwight D. Eisenhower.

Bert Lance, who resigned under fire for earlier banking dealings as head of the Office of Management and Budget in the first year of the Carter administration, also became associated with the tug-of-war for control of Financial General Bankshares.

In March 1978 the Securities and Exchange Commission (SEC) charged Lance and the Arab investors with attempting to secretly purchase, in violation of securities law, a controlling interest in the much-prized bank holding company. The SEC suit was settled with Lance and the others neither admitting nor denying any guilt.

Lance had purchased 13,000 shares in Financial General, investing more than $1.43 million of his money. Until January 1978, however, Lance had been deeply in debt as a result of outstanding multimillion-dollar loans to other banks.

Where did Lance suddenly find this money to invest? In January 1978, Lance sold 120,000 shares of the bank he owned, the National Bank of Georgia, to Saudi investor Ghaith Pharoan. At $20 a share, the sale gave Lance a hefty infusion of cash—$2,400,000. But it also prompted speculation, as *Business Week* pointed out, that "Pharoan was bailing him out to curry favor with President Carter."[4] Both Lance and Pharoan heatedly denied this. (As the new owner of the National Bank of Georgia, Pharoan inherited the $4.7 million loan to Carter's family peanut warehouse business.)

Still, the money Lance received was not enough to pay off all his debts. So where did Lance get the rest of his sudden liquidity? From a $3.5 million loan provided to Lance on the day he finished his deal with Pharoan—January 4, 1978. The loan came from the London-based Bank of Credit and Commerce International, an Arab-controlled bank whose president, Agha Hasan Abedi, had been one of the people trying to take over Financial General. Published financial records researched indicate that Lance was never asked to sign a note for the $3.5-million loan. Two years later, the loan had still not been repaid.

In August 1978, Lance gave an interview to the *Atlanta Journal and Constitution Magazine* in which he defended Arab investments in the United States as no different from other "multinational investments [which] have been a strong part of the American economy for a long time." He then suggested that the media's attention to Arab investments arose out of a particular reason: ''I don't know whether all the hurrah stems from the great Jewish ownership of the press or not.'' Several days later, Lance apologized on Atlanta television for his comments, but the nature of his apology belied a continued belief in the ''great Jewish ownership of the press'': ''I did not perceive this to be an offensive remark, and if, to the contrary, my statement offended anyone in any way whatsoever, I truly regret it.''[5]

Since 1978 Lance has continued to serve as an adviser and middleman for Arab and Middle East investors in the United States through his one-man unincorporated investment consulting company, the Lance Company, based in Calhoun, Georgia. According to financial sources, Lance has derived a substantial part of his income from this consulting.

The prominent Houston law firm, Vinson and Elkins, which handled Pharoan's December 1977 purchase of Lance's stock, was associated with an even greater political heavyweight, John B. Connally. The former governor of Texas who served as Nixon's secretary of the treasury had teamed up earlier in the year with Pharoan, whose enterprises did almost $2 billion a year in business, and another wealthy Saudi investor to purchase the controlling interest in the Main Bank of Houston. Connally and his law firm received so much Arab business that *Newsweek* wrote that the imposing six-foot-two-inch white-haired attorney was known as the ''top Arab money lawyer'' in Houston.[6]

Unlike most other members of the petrocorporate class, Connally and Lance had no direct ties to Arab governments or official Arab financial institutions. The fact that they became so deeply tied up with Arab oil money represented equally as much the natural diffusion of the enormous Arab wealth through the institutional organs of the American economy as it did the Arab desire to exploit these advisers' previous government service. Indeed, it was just a matter of time before handling Arab investments would become routine for thousands of financial advisers, lawyers, and bankers throughout the country.

But Connally soon demonstrated the subtle way in which oil money could engender goodwill. As a candidate for the Republican presidential nomination, Connally unveiled his own blueprint for Middle East peace on October 11, 1979, before the Washington Press Club. Though Connally was obviously making a bid to capture the limelight amid a thick pack of other candidates, the fact that he discussed the Middle East and the manner in which he did so reflected the political inroads petrodollars had made. In his nine-point plan, Connally called for Israeli withdrawal from all the occupied territories (except for "minor border rectifications"), dismantling of all Israeli settlements, Palestinian "self-determination," and Arab recognition of Israel. But the most controversial aspect of the plan was the link Connally saw between Israeli concessions and the willingness of Arab nations to "return to stable oil prices" and "forsake the oil weapon."

To many, the oil-for-Israel link was not only naive but also dangerous, fostering the illusion that only Israeli action stood in the way of lower Arab oil prices. Commented Felix G. Rohatyn, a widely respected New York investment banker who, the *New York Times* noted, had been a critic of various Israeli policies: "To think you can trade Israel for oil is totally impractical, in addition to being immoral."[7]

Others in the petrocorporate class were prepared to unabashedly exploit Arab sentiments about Israel and even about Jews in order to enrich themselves. Former Rhode Island Governor Philip Noel, an attorney who represented oil companies, and Richard Callahan, a former U.S. Drug Enforcement Administration officer, negotiated directly and indirectly with Syrian and Abu Dhabi officials on a special arrangement: in exchange for oil contracts to American oil brokers, a percentage of the revenues was to be "donated to a New York–based pro-Arab information center to counteract "Jewish bias" among the American public and thus circumvent the Foreign Agents Registration Act.[8] The scheme never materialized.

The most blatant—and frightening—demonstration of the depths to which a person would plunge to enrich himself was

revealed by former Vice-President Spiro T. Agnew in the spring
of 1976. "Zionist influences in the United States," he complained on NBC's "Today" show on the morning of May 11,
1976, "are dragging the U.S. into a rather disorganized approach in the Middle East" and have prevented the United States
from adopting an "evenhanded policy." When pressed by interviewer Barbara Walters to be more specific about the nature
of those "Zionist influences," the former vice-president replied
that they could be found in the "nationwide impact media."

In other interviews given on his coast-to-coast tour promoting his novel, *The Canfield Decision,* published in 1976,
Agnew seemed as obsessed as his novel's protagonist with the
manipulative power and cabals of the Jews. In one newspaper
interview Agnew was even more explicit than he had been on
television: the "Zionist lobby" has brought about a "disastrous
policy in the Middle East," Agnew asserted, adding, "I see no
reason why nearly half of the foreign aid this nation has to give
has to go to Israel, except for the influence of this Zionist lobby."
He claimed that the Jews owned or managed 50 percent of the
"national impact media"—which included the major wire services, pollsters, *Time* and *Newsweek,* the *New York Times,* the
Washington Post, and the *International Herald Tribune.* Specifically citing William Paley of CBS and Julian Goodman of NBC,
Agnew also spoke of "the tremendous" Jewish "voice," that
existed "not only in the media but in academic communities,
the financial communities, in the foundations, in all sorts of highly
visible and influential services that involve the public."[9]

Many observers found Agnew's fulminations against the
Jews almost inexplicable, given Agnew's presumably friendly
disposition toward Israel. William Safire, a *New York Times*
columnist and former speech writer for Agnew, attributed the
outbursts to Agnew's personal resentment of some of the Baltimore Jews who had turned state's evidence against him in court
several years earlier. A 1976 probe by the *Washington Post* of
Agnew's global business dealings concluded that the "picture of

Agnew as hired gun for Arab propagandists seems too simply drawn.''[10] The newspaper could find no evidence that Agnew was in the pocket of Arab officials.

However, in one little-noticed comment on his 1976 promotional tour, the former vice-president disclosed, perhaps unintentionally, the possible genesis of his views. Interviewed on the Merv Griffin television show, Agnew remarked that Saudi King Faisal, ''one of the most wonderful men I have ever met in my life, was so sympathetic and helpful to me, giving me the opportunity to get into some business matters there.''[11] Just *how* wonderful the king was, Agnew did not reveal on the Merv Griffin show or during his tour. But during 1976, Agnew was already involved extensively in acting as a middleman between Saudi government and foreign business interests. And according to financial documents, Agnew received an $80,000 commission for his role in one Saudi project alone: he had helped secure a $4 million construction contract with the Saudis for a Baltimore-based firm, the Atlantic International Corporation.[12]

Indeed, Agnew—who borrowed $200,000 from his friend, Frank Sinatra, after he resigned because of his self-acknowledged ''destitute'' status—was well on his way toward affluence as a result of his business dealings in the Arab world. Despite the nationwide editorial condemnation of Agnew, the more controversy he created, the more his Arab friends favored him.

Agnew had resigned from the vice-presidency on October 10, 1973, after pleading guilty to one count of income tax evasion. Once out of office, Agnew immediately became a middleman for Arab money. His gravitation toward the Arabs was not due entirely to the emergence of petrodollar riches; Agnew's enticement was rooted in earlier experience. In 1971 he had traveled to Iran as Nixon's emissary to participate in the festivities celebrating the 2,500th anniversary of the Persian Empire. Among the world leaders with whom Agnew roomed in the shah's $15-million tent extravaganza was King Faisal. Shortly thereafter the king, in a generous show of affection, provided Agnew with a diamond-studded gold-sheathed dagger; the crown prince of Saudi Arabia added to Agnew's treasury by giving him diamond and pearl jewelry. Not until April 1, 1974, after prodding by government officials, did Agnew hand the gifts over to the State Department, as required by law.

On that same day, April 1, 1974, Agnew initiated one of his first business ventures involving Arab oil money. He visited the offices of SRI International in Palo Alto, California, having arranged meetings with the research institute's top expert in synthetic protein production, Herb Stone. The assistants he brought along turned out to be the New York representatives of Aristotle Onassis, the billionaire Greek shipping magnate. Agnew, in fact, had teamed up with Onassis in an effort to sell Kuwait a petrochemical complex whose by-products could then be used to manufacture synthetic protein. In turn, the synthetic protein could be used as a substitute for the expensive organic protein in animal feed and even human food. A month later Agnew was on his way to Kuwait—one of many visits he would make to oil-rich Arab countries over the next several years—but the Kuwaitis did not accept the deal.

Later in the year, Agnew was given a much-publicized $100,000-a-year retainer by Indiana entrepreneur Walter Dilbeck for the explicit purpose of attracting Arab money for Dilbeck's projects such as hotel resorts. But Dilbeck soon claimed that Agnew had failed to bring in Arab oil money. Amid bitter recriminations by each, their relationship soured and was terminated in early 1975. Arab business deals did not seem to be going that well for Agnew.

During that time Agnew began working on the novel that was finally released in April 1976. The major theme of *The Canfield Decision* was the manipulative power of the Jews. In his cross-country media tour promoting the book, Agnew expanded on that theme. In the meantime, his business ventures and consulting in Saudi Arabia and other Arab countries began to net him hundreds of thousands of dollars in commissions.

In the years immediately following the petrodollar revolution, Agnew, for the most part, stood alone in the petrocorporate class in his outbursts against Israel and the Jews. But the growing number of influential, well-connected Americans who made their livelihoods from Arab money—and the occasional crass pandering to Arab hostility toward the Jewish community—symbolized the cultivating power of the petrodollar in the United States. It was not long before this influence would make itself felt in an extraordinary way at the highest levels of the U.S. government and Congress.

8

THE SUBCOMMITTEE AND THE SUPPRESSED REPORT

"How's the weather?"

"The weather? That's what you're calling about?"

*Telephone exhange between Senate
Foreign Relations Subcommittee
staffer in Washington, D.C., and the
U.S. Naval Station in Ras Tanura,
Saudi Arabia, 11,000 miles away*

This is the most shocking information I have ever seen in all my years at the agency," exclaimed the CIA oil and energy analyst, as he examined a draft of an investigative report prepared by the staff of the Subcommittee on Foreign Economic Policy of the Senate Foreign Relations Committee, on which I served. The investigation was so politically sensitive that the CIA analyst was discouraged by his superiors from consulting with the subcommittee staff. He did so anyway, after business hours, resulting in a major reassignment.

The information developed by the Subcommittee on Foreign Economic Policy was truly astonishing—showing, for example, that the Saudis' oft-mentioned link between oil production and American Middle East policy was a myth; that Aramco oil partners had mismanaged the Saudi oil fields; that oil company officials had claimed that the Saudis had more oil in the ground than they actually did so the companies could justify higher production rates; that Saudi officials were furious with Aramco at various times for causing potential structural damage to the oil reservoirs by overproduction; and that the 1973 Arab oil embargo was a cover-up for geologically and technically mandated oil production cutbacks.

However, most of these revelations—and all of the substantiation behind them—never saw the light of day. Though a sanitized version of the report was eventually released in April 1979, most of the shocking material had simply been suppressed. All quotes, names of Saudi officials and oil company executives, and dates and times of meetings were removed. The report was, short and simply, castrated. To this day the information has never been revealed to the American public because of the fierce pressure brought to bear on the subcommittee by Saudi leaders, Secretary of State Cyrus Vance, top State Department officials, and oil company executives. The episode marks the first time that Saudi Arabia was able to so thoroughly interfere with and impede the internal operations of an investigating congressional committee. Moreover, the entire five-year saga demonstrated the dangerous gaps in American intelligence concerning Saudi oil production and the unwillingness of the CIA and the Carter administration to challenge conventional wisdom.

In response to the excessive American dependence on foreign oil, President Nixon declared in 1974 a national program to make the United States energy self-sufficient by the end of the decade. His announcement was a cruel hoax: Project Independence existed in name only. By 1976, two years later, the OPEC countries controlled 85 percent of the total world crude oil exports, 52 percent of world production and 65 percent of world petroleum reserves. Saudi Arabia possessed more than 25 percent of the world's proven and probable reserves. And, since the cost of production for each barrel of oil to Saudi Arabia was only twenty-nine cents, it dampened the incentive for the major companies to invest in oil elsewhere. The United States remained as critically dependent on foreign oil as ever; one out of every two barrels of oil consumed by the American public was imported. By 1978, four years after the start of Project Independence, the United States relied on Saudi Arabia for 17 percent of all its imports.

OPEC ministerial meetings were becoming international media events as hundreds of journalists swarmed around, and the world helplessly watched as thirteen men decided the price of oil as if they were playing a board game. Every dollar by which they increased the price of oil automatically translated into hundreds of millions of dollars being sucked from the entire world economy. Like dangling marionettes, the less developed countries of the Third World and, to a lesser extent, the industrialized nations of the West were jerked around by a small clique of outside forces.

One OPEC meeting—awaited with unusually keen interest in the United States—was planned for mid-December 1976. Because Iranian oil reserves were much smaller than those of Saudi Arabia, the shah, knowing his production capability would peak and then start to decline in only a few years, wanted to maximize his revenues in the short term. Saudi Arabia, however, looked to the longer term, as it did not want to drive the price of oil high enough to force the West to develop alternative energy supplies. Nor did the Saudis want to jeopardize the value of their substantial investments in the West.

Prior to the December OPEC meeting, however, a typical conflict had erupted within OPEC, pitting the "hawks," led by Iran, against the "moderates," led by Saudi Arabia. The Shah demanded a 15 percent increase, and the Iraqis were calling for an even greater hike—25 percent. Sheikh Yamani, however, said his country would accept an increase of no more than 10 percent, but really hoped to limit the increase to just 5 percent.

Officials of the newly elected Carter administration, though they had not yet taken office, were very concerned about the deleterious effect that even a 10 percent increase would have on their economic recovery program. In addition, speculation was rising on Capitol Hill about whether the Saudis would make some type of conciliatory political gesture to the new presidential administration. Holding the price of oil would have been the perfect gift.

In the meeting held on December 15 in Doha, the capital

city of tiny Qatar, the bickering OPEC ministers failed to achieve a consensus regarding the price of oil. Saudi Arabia and the United Arab Emirates both announced that for the entire next year they would increase the price of oil by 5 percent, from $11.51 to $12.03 a barrel. The remaining oil producers, however, immediately scheduled a price hike of 10 percent, to $12.70, and further mandated an additional 5 percent hike on July 1, 1977. But Yamani, in a dramatic announcement, declared his government's unwillingness to allow the price hike set by the others. Therefore, he said, Saudi Arabia would keep the price down by sharply increasing its production to 9.8 million barrels per day by April 1—a significant boost over its annual ceiling of 8.5 million barrels per day.

Saudi Arabian officials also let it be known that they would increase oil production each quarter until they reached 11.8 million barrels per day by the end of 1977. The "new Saudi plans," reported the authoritative *Petroleum Intelligence Weekly,* "will mean an increase of nearly 3 million b / d" because Saudi production plans had previously called for a cutback of daily oil production to 7 million barrels a day.[1] Yamani also stated that progress on the Arab-Israeli dispute would affect his country's determination to hold prices down.[2]

Naturally, incoming Carter administration officials were extremely gratified by the Saudi declaration to increase production and keep the oil price fairly stable. To the new administration, which had campaigned so vigorously on its promise to revive the American economy, the Saudis had justly earned valuable political points. The IOU ledger was opened.

Ensconced in their cramped, sixth-floor offices in the dilapidated old Immigration Building, located on the Senate side of the Capitol, the staff of the Senate Subcommittee on Multinationals of the Senate Foreign Relations Committee had been carefully monitoring developments regarding Saudi oil production. Their interest was nothing new: in 1974 and 1975 the subcommittee, headed by Senator Frank Church, the populist Democrat from Idaho—who distrusted the revolving-door foreign policy elite who jumped from the business world to government and back to the business world—had produced the most extensive congressional investigation ever of the role played by multinational corporations in covertly influencing American foreign policy. The subcommittee had uncovered ITT's role in destabilizing the government of Chilean President Salvador Allende and had also revealed multimillion-dollar payoffs and bribes to foreign officials by Lockheed, Northrup, and other military exporters.

A substantial part of the subcommittee inquiry had also focused on the activities of the international oil companies from their earliest inception in the 1920s and 1930s, their operations in and relations with Middle East nations and the American government, and their role as accessories to the 1973 Arab oil embargo. Attesting to the voluminous amount of new information developed by the subcommittee was the fact that it spawned four successful books in the 1970s—three by British author Anthony Sampson, *The Arms Bazaar, The Seven Sisters,* and *The Sovereign State of ITT*; and one novel by Paul Erdmann, *The Crash of '79.*

Led by the intense but brilliant attorney Jerome Levinson, the nine-member subcommittee staff also explored, though somewhat tangentially, the physical condition of the Saudi oil fields. Part of the impetus for that particular line of inquiry had been provided by columnist Jack Anderson who had received information from internal oil company sources pointing to calamitous technical problems in the oil fields. Armed with congressional subpoenas, the subcommittee obtained oil company documents and appearances by oil company officials before the committee. Except for one official, the oil company men denied the existence of any acute technical problems. The one exception was William W. Messick, the chief reservoir engineer of SoCal and one of Aramco's top experts on the Saudi oil fields. In testimony given under oath, Messick admitted that Saudi oil fields had experienced precipitous drops in the reservoir pressure.[3]

In the Saudi oil fields, as well as in the older oil fields around the world, prudent maintenance requires that reservoir pressure be maintained above a certain desired level in order to push the oil to the surface. If pressure falls below the minimum levels, oil is lost and cannot be recovered except by prohibitively expensive recovery techniques. To keep pressure from dropping below the minimum level, water must be injected into the oil field to replace the removed oil. Messick revealed that the companies had failed to inject a sufficient amount of water into the reservoirs in 1973 because of technical problems. As a consequence, the reservoir pressures were dropping to unacceptable levels by October 1973, requiring some of the big oil fields to be shut down entirely.

Aramco had been "taken off the hook" by the Arab oil embargo, Messick told Senate investigators. He added, however, during the June 1974 Senate hearings, that Aramco had been able to increase water injection to desired levels since the latter part of 1973.[4] Other oil company officials denied that the oil fields were in disrepair to the extent Messick had alleged. Moreover, Senate investigators were unable to substantiate all of Jack Anderson's published allegations and his Senate testimony, nor could they corroborate the existence of incriminating oil company documents that the columnist claimed to have seen. There was a lot of smoke—but no gun.

The lifeline of the West clearly depended on the healthy state of the Saudi oil fields. Messick's testimony was unsettling—despite his assurances that the problems had been corrected. And Anderson's charges, though not proven, at least had been partially confirmed by Messick's testimony. Major questions thus lingered. Could the technical problems in the Saudi fields recur? Was the oil embargo actually a cover for technically required cutbacks? The staff continued to track oil developments.

Perhaps, however, Levinson's hunches were unfounded. Maybe Messick had indeed been wrong all along; maybe the other oil company officials were correct.

By December 1976, it appeared more likely that Levinson's suspicions had indeed been wrong. Yamani's declaration to increase Saudi production to 11.8 million barrels per day immediately following the breakup of the mid-December OPEC meeting indicated that Saudi Arabia could raise its oil output significantly and without any problem.

But in early 1977 jarring material came across Levinson's desk: the January 24 and February 14 issues of *Petroleum Intelligence Weekly (PIW)*, the widely respected newsletter that was known to have close links to both the Saudi government and the oil industry. *PIW* reported that in January the Saudis had been unable to raise production as they had promised; production had actually plummeted by 678,000 barrels per day. Why? The newsletter attributed the decline to a siege of "bad weather" in January so severe that it had prevented loadings of the oil tankers at the giant Saudi oil port of Ras Tanura. Although the report did not receive widespread media coverage in the United States, the absence of the increased Saudi production had immediate and far-reaching impact. It meant that the OPEC radicals would be able to raise the price of oil 10 percent.

To Levinson and other staff members, however, the *PIW* report of the "bad weather" seemed to be an unusually fortuitous occurrence for the Saudis: suddenly, with the Saudi inability to increase production, the shah of Iran was no longer threatening, OPEC became a cohesive force, the 10 percent price hike stuck, and Saudi Arabia was able to reap additional oil revenues. Yet the most important gains were the political dividends that Saudi Arabia earned without paying any costs: the kingdom had engendered goodwill in the United States, especially with the new President, for its supposedly courageous resistance to the avariciousness of the other OPEC ministers.

On March 16 President Carter declared that the Palestinians deserved a "homeland"—a position consistent with the President's emphasis on human rights. But coupled with information received several months later by a congressional committee that the Carter administration had authorized secret negotiations with the PLO by way of Saudi intermediaries—and that Carter was prepared to go beyond a "homeland" and accept an independent Palestinian state—the new position adopted by the President indicated that Carter may have cut a deal with the Saudis. When the President bade good-bye in June 1977 to his newly appointed ambassador to Saudi Arabia, John West, he had effusive praise for a country whose human rights record was dismal: "There has not been any nation in the world that has been more cooperative than Saudi Arabia."[5]

Again, perhaps Levinson was wrong, but this time the situation merited further verification. So he and another staff member drove to the U.S. Weather Service in Baltimore, about an hour's drive from the capital. Levinson asked the meteorologist whether the bureau's satellites had shown any storms for the previous five months in and around Saudi Arabia's main loading port. "What storms?" came the reply, "the weather patterns there have been perfectly normal." Somebody was terribly wrong, Levinson thought. Either the weather service or *PIW*—which, in effect, meant the Saudis or Aramco or both.

Levinson then arranged to see the logs of weather conditions that American flagships are required to keep and file with the U.S. government. These logs, which record weather conditions every six hours, are stored on computer printouts in Asheville, North Carolina. Levinson studied the logs of all the tankers that were anywhere near Ras Tanura at the time in question. They showed no unusual changes in waves, wind speed, or other weather conditions. The final confirmation that the *PIW* report had been totally erroneous was provided when Levinson persuaded the U.S. Navy, after much cajoling, to patch him through to its station in Ras Tanura, 11,000 miles away.

"How's the weather?" he asked when he was finally connected. The naval station operators were incredulous: "The weather? That's what you are calling about?" Levinson then asked them whether they had observed or monitored any particularly devastating storms in the past six months. "No," was the response, "we've seen no unusual weather." As if that information was not sobering enough, reports began to reach the sub-committee from reliable overseas sources connected to the Aramco partners that weather conditions were not the real cause of the Saudis' sudden inability to increase production. These same sources also stated their belief that Aramco had been forwarding disinformation.

What was really going on? Why were the Saudis unable to increase production after having promised resolutely to do so? If weather was not the real culprit, were the Saudis again covering up for technical problems in the oil fields? Was the American government aware of any deliberate attempt to pass on disinformation?

In further pursuing their questions, staff investigators conferred with the Central Intelligence Agency and the National Security Agency (NSA). If anyone could provide some answers, it was these two national intelligence-gathering agencies—the best in the world. But the mystery associated with the nonexistent "bad weather" in Saudi Arabia was soon to be transformed into a national security crisis as a result of the subcommittee's inquiry.

The staff went first to the supersecretive NSA, which, because of its ultrasensitive and covert intelligence-gathering methods, does not ordinarily brief congressional staffers or even congressmen. The agency, whose budget exceeds by billions of dollars that of the CIA, collects information by such unorthodox methods as electronic intercepts of telephone calls, cables, and microwave transmissions. But the agency wanted to patch up the strained relations with Senator Church that had developed out of Church's earlier investigations into abuses by the intelligence community. So the NSA gave the subcommittee staff a full briefing. The agency said it had been intercepting the coded messages, transmitted by a dummy Japanese company based in Ras Tanura. This company had been monitoring tankers at Ras Tanura and making detailed measurements, such as how deep the tankers were in the water—an indication of how much oil they carried. Based on this data, NSA experts estimated that Saudi production never exceeded 8.5 million barrels per day during the first quarter of 1977.

A visit to the CIA's headquarters across the Potomac in Langley, Virginia, was the next order of business for the subcommittee staff. At first the CIA—not knowing that the investigators had already been briefed by the NSA—denied that it or the NSA had any raw intelligence data about Saudi oil production. After a heated argument, the CIA officials finally conceded that they possessed material showing that Saudi Arabia had been producing an average of 9.4 million barrels per day during the first quarter of 1977—nearly a million barrels per day more than the NSA's estimate.

With a tight oil market, the absence or presence of an additional million barrels per day could make or break the American economy. In fact, two years later, in mid-1979, a 400,000-barrel-per-day shortfall—representing just 2 percent of the nation's oil imports—caused a national energy crisis resulting in mass panic and long gas lines.

But after further scrutinizing the raw data provided by the CIA, Levinson discovered that the CIA had inflated the Saudi production figures. Oil for export had been lumped together with oil for use in Saudi Arabia itself and with oil produced in the neutral zone, a border area that Saudi Arabia shared with Kuwait.

A Pandora's box was opened. Had the CIA or Aramco or even the Saudi government deliberately misrepresented Saudi production levels? If so, why? Why was there such a discrepancy in the data? Did Saudi Arabia have major technical problems? Was the Carter administration afraid that public exposure of the Saudis' true oil limitations would undercut the public relations representation of Saudi Arabia as America's best friend in the Middle East? Did the disagreement between the CIA and NSA belie a critical lack of intelligence-gathering capability in Saudi Arabia? No one could properly explain the precipitous oil production drop, the nonexistent "bad weather," and the stark disagreement between the CIA and NSA.

Then, suddenly, the coup de grace occurred. Two major fires erupted in Saudi pipelines and pumping stations in May and June 1977, causing $60 million of damage and resulting in a drastic cut of over 5 million barrels per day in oil production. Moreover, critical operatives within the intelligence community were not able to get full details of the second fire until two weeks after it happened. It was becoming abundantly clear to Senator Church, who was being briefed continuously of his staff's findings, that the U.S. executive branch had a gaping hole in its intelligence of, and policy toward, one of the most important external political and economic issues affecting the country's future—the supply of Saudi oil.

In spite of the brutal oil shock of 1973, government experts and policymakers still looked on Saudi Arabia as the only oil-producing nation that could and would increase production and capacity to match the insatiable Western appetite. This meant that Saudi Arabia would be in a position to pump oil long after its OPEC counterparts peaked. Moreover, Aramco—the best possible authority—had repeatedly promised to American policymakers that Saudi Arabia would increase its oil production capacity by 50 percent, to 15 million barrels per day by the early 1980s. During the 1974 Multinational Subcommittee hearings, some Aramco executives in fact had talked optimistically of a 20-million-barrel-per-day production target. As late as May 1977, Frank Jungers, Aramco's head, declared that "we're headed for 16 million barrels a day of capacity by 1982." Another Aramco official, Joseph Story, was even more emphatic about the critical role Saudi Arabia would play: Saudi Arabia "is the only country that can meet the incremental energy import requirements of the United States and the world over the next decade or so."[6]

The CIA, the Department of Defense, the Department of Energy, and the Department of State all had incorporated the Aramco predictions—Aramco was often the only source of such information—into their strategic planning for the country. In 1976, for example, the State Department, in an internal report, predicted that Saudi capacity would be increased to 16 million barrels per day by 1983.[7] The CIA was predicting that by 1985 the world would need 19 to 23 million barrels of Saudi oil per day. Underpinning these estimates was the implicit assumption that the United States would have to coax the Saudis to raise production to this level, presumably by political inducements. In short, a great deal depended on the availability of Saudi oil. America's foreign policy and national security, arms sales, the Rapid Deployment Force, concessions on the Arab-Israeli dispute—all largely hinged on how much oil the Saudis could and would produce. The comment of one unidentified high-ranking American government official, quoted in an article in the *Washington Post* by Richard Harwood, seemed to summarize the prevailing belief: "If the United States will get from Israel a Palestinian homeland, you can get all the oil you want. If you don't get a settlement [an Israeli withdrawal from the territories captured in the June 1967 war], I don't know what will happen."[8]

As no one in the executive branch seemed truly concerned about the problem of Saudi oil fields, Church wanted the issue fully investigated and brought to the attention of the American public. He authorized a preliminary investigation to determine whether any technical problems had impeded oil production in Saudi Arabia, to learn the true state of oil production in the kingdom, and to find out what hanky-panky, if any, Aramco had engaged in, in its zeal to extract as much oil as possible before it was nationalized.

The forty-five-year-old Levinson, due to an almost fatal heart attack, was forced to leave the staff; James Thessin, a thirty-two-year-old attorney with a computerlike mind for plowing through and digesting complex data, was brought over from the Federal Trade Commission, to carry out the investigation. Having studied at one time in a seminary, Thessin was guided by a deep commitment to fairness, an attribute that made him fastidious, thoroughly conscientious, and exceedingly cautious. If he put his name on a report, it was correct. Also participating in the inquiry, but to a lesser degree, were thirty-four-year-old Karin

Lissakers, the staff director of the subcommittee, and myself. (I joined the staff of the subcommittee in December 1977.)

Immediately, however, the subcommittee was confronted with a major problem, the same one that had prevented the government from confirming the truthfulness of Aramco's assurances and its presentations to other branches of American government: only the oil companies had possession of the highly sensitive internal documents, and they were not going to turn them over to some congressional subcommittee—or anyone else, for that matter. From the vantage point of the oil companies it was easy to see why. Besides proprietary concerns and fear of offending their Saudi patrons, it would have been absolutely foolish to release evidence of oil field mismanagement or reckless business practices. By late 1977, the chairman of Aramco, John J. Kelberer had already staked out a stiff public position in response to allegations of shoddy oil production techniques and practices. Responding to an extensive *New York Times* article written on December 25, 1977, by Pulitzer Prize–winning reporter Seymour Hersh about Saudi oil production problems, Kelberer wrote in a letter to the editor: ''Aramco follows first class oil field practices in all of its operations.'' [9]

One of the major weaknesses of American national security policy since World War II had been that the government essentially entrusted the major oil companies with formulating national energy policy. The government had no way of confirming oil company reports, and apparently could not accept the fact that the foremost allegiance of the companies was to their profits and to their own survival, not necessarily to the national interest.

So, in spite of Kelberer's statement and the complacent attitude adopted by the CIA, the stakes were too great to allow an investigation to rest on the self-serving testimony and statements of the oil company officials and other branches of government relying on Aramco data. Therefore, the only way to obtain the necessary data was to subpoena internal oil company records. In a matter as strategically important as this, it was vital to get at the truth even if this meant offending Saudi Arabia or the oil companies. Yet, within the subcommittee, there was immediate resistance to the idea of subpoenaing the oil companies; the resistance, however, came from a particularly unlikely source—Senator Jacob Javits of New York.

Known as a Senate workaholic, Javits, who was first elected to the Senate in 1954, was the ranking Republican on the five-member subcommittee. He had an ingenious mind and unrelenting drive to stamp his imprint on legislation that passed through his hands. He had been a Republican all of his life, but was considered one of the most liberal members of the Senate, having been at the forefront of the battles for civil rights and congressional oversight over presidential war-making authority. His ratings from Americans for Democratic Action, a Democratic liberal interest group in Washington, had always been among the highest in the Senate.

Javits was a Jew, too, and represented a state with one of the largest concentrations of Jews in the United States. Javits became one of Israel's most enthusiastic supporters on Capitol Hill. For example, whenever markup time on foreign aid legislation came around, Javits would invariably find a way to increase American aid to the Jewish state. From the perspective of Arab countries, Javits was considered squarely in the camp of the pro-Israel lobby. It seemed, therefore, that investigating the oil companies and their ties to the Arab governments would have been not only consistent with Javits's philosophy but also politically smart.

But the seventy-four-year-old Javits had another philosophical streak in him, one that he managed to keep under wraps. In representing the state of New York, Javits acted to protect the interests of his most powerful constituents: big business and the international banks. Believing strongly in the efficiency of corporate capitalism, he sincerely equated the interests of multinationals with the American national interest. Javits saw multinational corporations as the major organs of our economic system, and many of those corporations—which were heavily involved in recycling petrodollars—were headquartered in New York City. Only once before had Javits been confronted so starkly with the contradictory strains of his beliefs: in the early 1970s Javits—despite his seniority—elected to stay off Frank Church's Subcommittee on Multinationals, according to subcommittee officials and a colleague of Javits, because Javits was not willing to participate in an investigation of corruption on the part of the multinational corporations.

Despite his earlier refusal to face a choice of defending or investigating the multinationals, Javits exhibited no such hesitation this time around: he fiercely protected the oil companies and their Saudi benefactors. He initially opposed issuing the subpoenas to the oil companies. The inquiry, he maintained, would also be an unnecessary embarrassment to the Saudis.

The categorical need for issuing the subpoenas was soon to be demonstrated by subcommittee staff members. Thessin and Lissakers had flown to Oregon to meet with Frank Jungers. The former chairman of Aramco admitted the existence of technical problems in the oil fields, but he was not specific enough. The most critical evidence would soon be provided by a most surprising source, W. Jones McQuinn, who had been a vice-president of SoCal and a director of Aramco. In vivid detail, McQuinn revealed to Thessin and Lissakers Aramco's deficiencies in managing the oil fields, the imposition of stringent conditions on various oil fields by the Saudis, and the absence of a commitment by the Saudis to reinvest in expanded oil facilities.

Confronted with this new information, coupled with a staff recapitulation of all other subcommittee findings, Javits backed down and admitted the need for the subpoenas. In late August 1978, congressional subpoenas—the most powerful type of legal subpoena—were issued to SoCal and Exxon, the two largest members of the Aramco consortium, and the two that had been responsible for producing the required internal technical assessments of the oil fields.

The oil companies were stunned. Besides Saudi pressure on them not to comply, they had to protect their own self-interest. Richard Keresey, a counsel for Exxon, was dispatched with two other oil company officials to meet with Thessin, Lissakers, and myself. During the tense three-hour meeting in the subcommittee's offices, Keresey, seated at the head of a rectangular oak table, decried the "fishing expedition." "We can't possibly respond to your subpoenas," he declared.

Keresey, who four years before had contended with the earlier subcommittee investigation led by Levinson, demanded that the subpoenas be redrawn and limited in their scope. But Thessin would not budge. He was not going to trust the oil companies to be the arbiter of relevant information. The final company defense was that the documents were Saudi property and that Exxon could not disclose proprietary material belonging to another government. That rationale would be repeatedly invoked throughout the investigation. In truth, however, the documents were wholly owned and possessed by the American companies. As McQuinn had pointed out earlier to Thessin and Lissakers, the technical assessments had been prepared on behalf of and presented to the American shareholders. The companies' protest against providing proprietary material of another country rang especially hollow: these same companies never hesitated at the time of the 1973 oil embargo to comply with King Faisal's demand for information as to where the American military got its oil supplies. In order to expand the embargo, Faisal demanded and received sensitive information from the oil companies on oil shipments to the U.S. Navy.

Keresey and his two colleagues stormed out of the subcommittee offices. Grudgingly and slowly, the oil companies began to send documents to the subcommittee. But when Thessin started examining the materials, he discovered that many documents, pages, and sections had been censored, deleted, or simply removed. Phone calls and additional meetings with the company officials became an exercise in futility. It was clear that a stonewalling action had begun, like the one SoCal had attempted earlier in response to a subpoena issued by the Multinational Subcommittee in 1974. At that time, SoCal produced just one document, in an attempt to force a showdown and show the other companies how to deal with the subcommittee. But they did not realize how determined Levinson and Church were; when faced with the threat of contempt proceedings, SoCal backed down. Similarly, after Thessin warned Exxon officials that continued withholding of documents would lead to proceedings charging them with contempt of Congress, they opened up virtually all their files to the subcommittee staff.

In their newfound zeal to be scrupulous, they actually flooded the subcommittee with ream after ream, box after box, of documents, ultimately pouring in over 100,000 pages, many of which were totally irrelevant. Methodically, Thessin sifted through every one of the documents.

Meanwhile, Church, having taken on the oil companies and America's corporate elite once before, was acutely aware of the array of powerful forces lining up against the report. He pressed the staff to complete its investigation before it "was too late." Working around the clock seven days a week, Thessin—at his frustrating but necessarily meticulous pace—produced a startling 130-page report.

But it was too late. The "full court press" by the oil companies and the Saudis had begun to take effect. Javits ordered that a copy of the report be forwarded to the State Department, where the Bureau of Near East Affairs studied it like a talmudic tract. Portions of the report were transmitted to the American embassy in Saudi Arabia. When informed about the investigation and report, oil minister Sheikh Yamani informed the American ambassador, John West, that Saudi Arabia would not tolerate the material being made public. These documents, he said, were the property of Saudi Arabia, and their release constituted an invasion of national privacy. Immediately, the embassy cabled back to the State Department, relaying Yamani's concerns. In addition, a copy of the report was made available to the oil companies—though by whom is still in question—an act akin to giving a preliminary grand jury investigation report to the subjects about to be indicted.

The subcommittee then scheduled a momentous vote on whether to release the report. Though Senate rules had forced Church to leave the subcommittee when he ascended to the chairmanship of the Foreign Relations Committee, the changes looked promising. The new chairman of the subcommittee was Senator Paul Sarbanes, a progressive, liberal, first-term Democrat from Maryland who was as firmly committed to the need for the investigation as Church had been. The other Democrats on the subcommittee were former presidential candidate George McGovern of South Dakota and Senator Joseph Biden of Delaware, both of whom shared the liberals' distrust of unnecessary governmental secrecy. Both were considered supporters of Israel, though Biden was an especially strong advocate of closer American-Israeli relations compared to McGovern who did not hesitate to criticize Israeli actions that he felt did not aid in the cause of peace. Besides Javits, the only other Republican was Senator Richard Lugar of Indiana, a cautious, moderate man who—despite being known as Richard Nixon's "favorite mayor"—was developing a reputation for independence in the Republican party. Held in the small but elegant anteroom next to the Foreign Relations Committee hearing room in the Capitol, the executive session was closed to the public, not because of national security, but because of the political embarrassment that might ensue if the proceedings were made available to the press.

Invited to make a presentation was Joseph Twinam, the forty-five-year-old deputy assistant secretary of state for Near East and South Asian affairs. With the exception of a short stint in Amsterdam, all of Twinam's overseas assignments had been in Arab nations such as Lebanon and Saudi Arabia. A soft-spoken man, Twinam rose very fast in the State Department bureaucracy, becoming the first resident U.S. ambassador to Bahrain in 1973 and then, in 1976, the deputy director of Arabian Peninsula affairs.

If the report were released, said Twinam, the Saudis would not react immediately. But eventually, Saudi displeasure would be felt in a whole range of matters vital to the United States, such as oil prices and Saudi dollar holdings, which at the time amounted to some $25 billion. A forceful discussion followed in which Javits argued strongly against the report's release while Sarbanes and Church—who had shown up despite his non-voting status—made strong pro-report pitches, urging their colleagues to get this vital information into the public domain. The roll call was read by the committee clerk:

"Mr. Sarbanes?"
"Yes."
"Mr. McGovern?"
"Yes" (by proxy).
"Mr. Biden?"
"No."
"Mr. Javits?"
"No."
"Mr. Lugar?"
"No."
"The nays have it, three to two."

The report was killed. Church and Sarbanes were shocked. The subcommittee staff was in a state of disbelief. But the outcome, unknown to us at the time, had been anything but a surprise. For the days preceding the vote, members of the subcommittee had been warned by oil company and State Department officials that the senators would be held responsible for the "consequences" if the report were released. Clifton Garvin, Exxon's head, called at least one senator and asked him to stop the report. The matter was deemed of such great national importance that Secretary of State Cyrus Vance became actively involved in trying to suppress the report. He telephoned members of the subcommittee, warning them that release of the study would imperil U.S.–Saudi relations.

With each phone call, the ante was raised, Church recalled. "Release of this report would be an irritant to U.S.–Saudi relations," Vance said to Church, adding, "The Saudis will look upon this as interference in their sovereign rights.''[10] In later phone calls, Church was told he would have to shoulder responsibility for imperiling the Egyptian-Israeli negotiations—a likely repercussion, Vance said, if the report was made public. Lugar was won over. And Biden—representing Delaware where thousands of firms are incorporated, including Aramco—wilted, too.

The project, however, was not entirely dead. For even Javits realized the overriding national interest of getting at least some of the information into the public arena. Moreover, Jack Anderson had written an embarrassing column decrying the shielding of Saudi secrets by the ''oil moguls'' on the Senate subcommittee. Javits's hand had been forced. Therefore, he agreed to the release of a sanitized version, one that would not offend the Saudis and would protect the oil companies.

The subcommittee staff produced a series of drafts, but each one was progressively truncated as a result of the unprecedented editorial involvement of State Department Arabists who meticulously scrutinized every word. Javits, the man who was responsible for the most important piece of legislation delineating the separation of powers—the War Powers Resolution—had authorized State Department control over a clear Senate prerogative.

At one point, an investment banker from Lehman Brothers, a former State Department official who had no special expertise at all in the energy area, was brought in, unbeknownst to Sarbanes and Church, to ''edit the report.'' Much of his editing involved removing or changing passages that, according to a committee memorandum, ''put Saudi Arabia or Aramco in an unnecessarily unfavorable light.''

Ultimately the report was reduced to thirty-seven pages.[11] On Saturday, April 14, 1979, it was released. Evidence of Saudi perception of substantial Aramco mismanagement, conflicts between Aramco and Saudi officials regarding appropriate levels of oil production, bitter accusations by Saudi officials that Aramco had caused irreparable damage by overproduction, and deceptive Aramco accounting of Saudi oil reserves and new discoveries—all had been removed. Additional historical evidence that severe pressure problems in certain Saudi oil fields had necessitated a severe production cutback in October and November 1973—the same time as the Arab oil embargo—also had been deleted. Ironically, the original report would have added immeasurable strength to the repeated Saudi requests to the West to conserve oil; the report would have confirmed that Saudi Arabia would never be willing to meet high levels of Western consumption.

The true story of the Saudi oil fields has never been told to the American public. Moreover, in an effort to further delegitimize the report's authoritativeness, all quotation marks were deleted, as were the names of oil company and Saudi officials. The most salient information in the final version was that the Saudis—for economic, geological, and conservationist reasons—were never going to produce more than 12 million barrels per day. But the credibility of the report was radically diminished by one glaring omission: nowhere in the final version is it stated that the report was drawn from 20,000 subpoenaed oil company documents.

The success of the Saudis in getting the subcommittee report censored was unprecedented in the operations of Congress.[12] Never before had a foreign power been able to interfere with an investigating committee. Within the next year, however, the Saudis attempted to exert their leverage on an even more surprising target: the media.

9

THE FATE OF A PRINCESS

The film was so sensitive that it might endanger the well-being of several hundred Alabamians working in Saudi Arabia and other Arab countries.

*Spokesman for Alabama
public television explaining
why* Death of a Princess
would not be shown.

Like a sudden desert sandstorm, the political tempest caught everyone off guard and quickly spread across four continents. Every day banner newspaper headlines ominously warned of dire political consequences for the West, such as a cutoff of Saudi oil or a trade embargo. In London, Amsterdam, Stockholm, Sydney, and Washington, government leaders labored furiously to contain the political damage and fallout. What had induced a state of panic among the leaders of the free world? A two-hour film about Saudi Arabia.

Entitled *Death of a Princess,* the film told the story of the life and death of a Saudi princess who, in defiance of Saudi law and ritual, eloped with a Saudi commoner instead of marrying the man her family had selected for her. Returning to Saudi Arabia, she confessed to having committed adultery. For that crime, the princess and her lover were sentenced to be executed—for violation of Islamic law—by the princess's grandfather, Prince Mohammed bin Abdul-Aziz, the oldest surviving son of Ibn Saud, the founder of Saudi Arabia. The uproar surrounding the television broadcast of the film in the United States marked the first demonstration of the widespread fear and goodwill Saudi Arabia had instilled among different components of the American population.

The $430,000 film was the product of a collaborative effort between the Boston public television station WGBH and Britain's Associated Television Corporation (ATC). The film presented a dramatic reconstruction of the events leading up to an actual 1977 incident in which a nineteen-year-old Saudi princess and her lover were publicly executed for adultery—she was shot and he was beheaded—in a parking lot in the port city of Jidda.

Having been given indisputable information by a source close to the royal family, filmmaker Anthony Thomas researched the motion picture for two years. He went to Saudi Arabia, but found no Saudi official or member of the royal family willing to provide authoritative information about the case. However, Thomas did find many conflicting accounts of the princess's death, each apparently colored by the highly polarized cultural-political prism. Unable to persuade anyone to appear on camera, Thomas used actors to make a dramatized documentary of his efforts as an inquiring journalist trying to ascertain what had actually occurred in Saudi Arabia. His screenplay, in fact, included verbatim conversations taken from his taped interviews.

Though the film never referred to Saudi Arabia by name ("Arabia" was used), there could be no doubt as to where the events had taken place. The film, however, went far beyond the princess's execution, providing a rare glimpse into Saudi society, especially the internal contradictions of a strict fundamentalist Islamic society caught up in a modern world. In the words of David Fanning, the film's producer, *Death of a Princess* was the "story of the Arab predicament—being pulled in three different ways: the desert, the modern world, and radical politics."[1]

The film touched on potential instability in the kingdom, referring to an unsuccessful coup by air force officers, many of whom were subsequently executed. And in one of the final scenes, the Saudi royal family is bitterly criticized by a young Saudi woman: "These people pervert Islam, they use Islam. They scare people to death with barbarous, illegal punishments. This is not Islam. . . . Islam is humane, gracious and forgiving."

On the evening of April 11, 1980, British television broadcast the film to an audience of about eight million Britons. Even before the broadcast, Saudi officials, who had been given an opportunity to preview the film, were livid. They were simply not going to tolerate such an unflattering portrait. The royal family saw the film as an attack on them, and by extension—owing to the logic of a family understandably obsessed with maintaining itself in power—as an attack on the Islamic world.

In Saudi Arabia, Foreign Minister Saud angrily summoned the British chargé d'affaires and demanded that the film not be shown. Immediately preceding and following the film, infuriated Saudi officials issued veiled threats linking their anger to a possible oil cutoff and a halt to importing of British goods. Either course of action would have hit Britain's national economic jugular: by 1980 Saudi Arabia had become Britain's eighth largest market, purchasing more than 2 billion dollars' worth of British goods and services. Saudi Arabia also supplied Britain with 12 percent of its oil imports. A raw nerve had been hit with British banks, industrial exporters, and defense manufacturers.[2]

Immediately after the broadcast, the British Foreign Office tried to appease Saudi anger with a public apology: "We profoundly regret any offence which the program may have caused in Saudi Arabia."[3] Moreover, British Foreign Secretary Lord Carrington and his deputy Ian Gilmour, both sent personal messages to King Khalid apologizing for the "understandable offence . . . this particular television film had caused to the Royal Family in Saudi Arabia, and other Saudis and Muslims everywhere."[4] (Carrington's apologies to Saudi Arabia were all too reminiscent, some members of Parliament felt, of an apology issued by Carrington's predecessor, David Owen, in 1977 to Saudi Arabia for a statement by a British official who had criticized the actual execution of the Saudi princess and her lover.)

Two days later British Trade Secretary John Nott, who had returned not long before from a visit to Saudi Arabia designed to promote British exports, asked derogatorily and rhetorically at a public meeting whether the producers of *Death of a Princess* had been on an "ego trip . . . knowing as they did that it would cause the gravest offence to the country in question to which we are doing our very best to increase our exports."[5] Every day a flurry of diplomatic exchanges between the two countries failed to contain the rage in the kingdom. One Saudi newspaper warned unambiguously that the British government "will not be allowed to escape without paying the price."[6] The Saudi newspapers fulminated at the "Zionists" in the West who were allegedly orchestrating a campaign against the Saudis.

The Saudi reaction was swift. The British ambassador was expelled. Declared the official Saudi press agency, the expulsion was due to "the British government's negative attitude toward the screening of the shameful film."[7] A planned trip to the kingdom led by secretary of state for defense Francis Pym was canceled. British landing rights for the supersonic Concorde were postponed as was a state visit to Britain by King Fahd. Members of the Saudi royal family were ordered home from England. But the most chilling Saudi announcement had nothing to do with any further rupturing of diplomatic ties. Instead, the Saudis dangled the threat of economic reprisals.

Quoted by the Associated Press and given prominent news coverage around the world, the Saudi spokesman, Information Minister Mohammed Abdu Yamani, declared that his country was "giving serious consideration to the economic relations between the Kingdom and Great Britain, particularly with regard to the work of British companies in Saudi Arabia."[8] Several weeks later the Associated Press reported that the American multinational construction engineering firm of Ralph Parsons, headquartered in Pasadena, California, had been ordered to boycott British subcontractors on projects at the city of Yanbu' by Saudi officials.[9] The official who issued the order was the head of the Royal Commission for Jubail and Yanbu', Farouk Al Akhdar, the same person who five years before had traversed the United States wielding the "money" weapon before the American business community. But Yamani's sanction transpired well after the controversy was over in the United States. There was no immediate evidence of any retaliatory economic steps imposed against British industry.[10]

Neither had any immediate trade reprisals been taken against various other countries, such as the Netherlands, despite Saudi warnings to their respective governments if *Death of a Princess* was broadcast. The governments of Italy, France, West Germany, and Australia, however, were not willing to incur Saudi anger: *Princess* was blocked from being shown in these countries. In Sweden, a group fronting for Saudi interests, according to PBS (Public Broadcasting System) sources, bought the rights to the film and suppressed its broadcast.[11] It seemed, however, that the Saudi threats were being made with the ulterior purpose of influencing a much more prized target than Britain or any other country: the United States.

Almost as soon as the film had been broadcast on British television, the focus of media attention shifted across the Atlantic. Would the Carter administration succumb to Saudi pressure and try to stop the film? Would the royal family issue threats or exact any punishments? American newspaper reporters began questioning Public Broadcasting System officials about their plans. But at their studios in the fortresslike L'Enfant Plaza section of downtown Washington, PBS officials staked out a tough position, publicly vowing as early as April 11 that *Death of a Princess* would be broadcast as scheduled. Eleven days later, when some of the fallout from the British episode was just beginning to materialize, PBS officials again stiffened their iron resolve and proclaimed that sanctions taken or threatened by the Saudis against the British would not affect their determination to broadcast *Death of a Princess* on May 12. "We're going to run it, period," declared Barry Chase, director of news and public affairs programming for PBS.[12]

Throughout the entire tense four-week period preceding the

broadcast, during which time a whirlwind of diplomatic pressure and political intrigue surrounded the upcoming PBS broadcast, Chase and Larry Grossman, the president of PBS, maintained that no pressure had been exerted on them, directly or indirectly, by the Saudis, nor had they wavered in their determination to show the film. In fact, however, PBS came perilously close to canceling the broadcast after a prominent Saudi official had issued oblique warnings to them, and after a PBS official adopted a startling position.

Although Grossman and Chase had meticulously followed the events as they unfolded in England and elsewhere, they had become painfully aware of the stakes at hand through a much less circuitous route: a direct Saudi pipeline to their offices. Using Jo Franklin (now Jo Franklin-Trout), a producer of the McNeil-Lehrer television show, as an intermediary, the Saudi ambassador to the United States advanced several explicit warnings to the two PBS officials. Producer Franklin-Trout served as the conduit for the Saudi messages because she had cultivated a good relationship with the Saudi ambassador as a result of her involvement in Middle East issues for McNeil-Lehrer. Franklin relayed the messages to Chase because of her concern that, according to Chase, "we not fly blind" about the possible consequences ensuing from broadcasting the film. As recalled by Chase, Saudi Ambassador Sheikh Faisal Alhegelan sent messages via Franklin that distinctly mentioned "possible repercussions" in the areas of "oil supplies" and "defense" and the possible recall of the ambassador.

Meanwhile, State Department spokesmen publicly confirmed that Saudi officials had substantial contact with American officials at Foggy Bottom, and announced in deceptively innocent sounding State Department language, "We have listened to their concerns and are checking into the situation."[13] And in a ploy designed both to ensure that PBS was intimately aware that the government was monitoring the situation and to avoid any accusation of overt pressure, a Saudi Arabian desk officer telephoned PBS to ascertain conspicuously how many local stations would be airing the film. Yet, State Department officials were extremely eager to have the film killed, "but," admitted State Department spokesman Jack Touhey, "legally there are limits as to what we can do." [14]

Publicly, PBS was still putting up a brave front. A spokeswoman declared on April 24, "We do intend to air the program on May 12." But behind the scenes, the story was much different. Serious thought was being given to canceling the broadcast. Moreover, in an effort to keep his options open, Grossman pulled a $24,000 advertisement out of *TV Guide* promoting *Death of a Princess*. The pressure continued to mount. According to PBS sources, a syndicate of anonymous American businessmen offered through an intermediary to purchase the "rights" to the film for a fee in excess of $10 million.

The second ranking Republican on the Senate Foreign Relations Committee, Charles H. Percy—a leading supporter of business interests on Capitol Hill—informed President Carter of his strong objections to the showing of the film and proffered his assessment that other members of Congress generally felt the same way. Percy, the former whiz kid who rose to the helm of Bell & Howell at the age of twenty-nine, told President Carter, according to an aide, that "there would be support in Congress should the President make a determination that the showing of the program would not be in the national interest."

This was not the first time that Percy had felt compelled to promote the views of big banking and the Arab oil exporters. In 1975 he had helped to thwart the efforts of the Senate Foreign Relations Subcommittee on Multinationals to determine the size of rapidly growing Arab bank deposits in American banks and the dangers attending to a potential sudden withdrawal. After having been personally warned by Kuwait Minister of Finance al-Atiqui that Kuwait would pull its funds out of American banks if there was any Senate disclosure of its investment position, Percy startled some of his Senate colleagues when he openly issued a stern caveat to the American Jews: "If Saudi Arabia and Kuwait withdraw their bank deposits, the biggest single loser would be the city of New York and I would say the American Jewish community, centered in New York, would be the largest loser." [15]

Even Percy's daughter became involved. Sharon Percy Rockefeller, wife of West Virginia Governor Jay Rockefeller, and a member of the board of the Corporation for Public Broadcasting, which oversees some of the funding for PBS, met with the head of CPB, Robben W. Fleming, to try to kill the film. "These are not normal times," she said after the meeting, "and the climate this [program] might be shown in is extremely delicate. We can hardly afford any adverse reaction." [16]

The former chairman of the Joint Chiefs of Staff, Admiral Thomas H. Moorer, also became embroiled in the affair, lobbying board members of CPB in an effort to squelch *Princess*. In an interview with Sanford Ungar on "All Things Considered" on National Public Radio, one of several interviews Moorer gave, he explained his position. Moorer was asked whether the United States ought to succumb to Saudi pressure to kill the show. The admiral, a director of Texaco (one of the four Aramco companies) and Fairchild Industries, a huge defense aerospace company, responded bluntly: "See you down on the gas line!" At another point in the interview, Moorer added an apocalyptic dimension to the debate when asked whether the United States ought to accept the Saudi opinion that the film was offensive. Said Moorer, "You can't stand on principle in the face of blowing up the free world." [17] When I asked Moorer four years later whether he really believed that the world would "blow up" if the film was shown, he said, "Sure, we just can't alienate our friends who have all this oil." [18]

On Capitol Hill the hysteria appeared contagious as both the chairman of the House Foreign Affairs Committee, Clement Zablocki of Wisconsin, and the ranking Republican on that committee, William Broomfield of Michigan, both attacked PBS for its scheduled broadcast of the movie. On the floor of the House of Representatives, Texas Republican Ron Paul lashed out at WGBH, the Boston station that had co-produced the film, threatening to terminate congressional funds to PBS and called the broadcast the "most serious diplomatic blunder in years." Three other members of the Lone Star State's congressional delegation also chastised PBS on the House floor.

The brunt of the public external pressure on PBS came from a swift, almost fatal one-two punch delivered by both Mobil Oil, a major PBS underwriter, and the State Department. In an advertisement on the Op-Ed page of the *New York Times* on May 8, 1980, Mobil unleashed a bitter attack on the film, PBS, and the media, claiming that the film "jeopardizes U.S. relations" with Saudi Arabia. Under the headline "A New Fairy Tale," Mobil—which had contributed $2.3 million in 1979 to PBS programming—cloaked its objections in a rather novel interpretation of the nation's Constitution, reducing the inalienable First Amendment rights to the level of "privileges" that could be taken away whimsically. The ad stated, "We all know that in the United States the Constitution guarantees a free and unfettered press. However, implicit in that guarantee is the obligation on the part of the press to be responsible. Clearly the people of the United States have the right to expect that the media will not abuse its privilege. The public will have to decide whether a 'free press' is acting responsibly." Mobil also assumed the role of arbiter of national interest when it urged PBS officials to "review its decision to run this film, and exercise responsible judgment in the light of what is the best interest of the U.S."

At midday on Thursday, May 8, Acting Secretary of State Warren Christopher forwarded to PBS a copy of a letter he had received from the Saudi ambassador. Compared with earlier threats made to the British and those communicated to PBS by the McNeil-Lehrer producer, the Saudi letter was relatively mild, expressing only "concern" about many "inaccuracies, distortions, and falsehoods of the film." The day before, however, Ambassador Alhegelan sounded a more conspiratorial tone: "The timing of the showing of the film in the present period makes it clear that it is part of the continuing and recent stepped up effort to undermine the international significant relationship between the United States and Saudi Arabia."[19]

Christopher's transmission of the Saudi ambassador's letter to PBS represented the culmination of two weeks of heated exchanges between the Saudi government and the State Department. According to State Department sources, the Saudis had in fact prepared a letter attacking the film that was to be signed by the State Department and then sent to PBS. The State Department agreed only to pass on the admonition from Alhegelan. The legal adviser's office was then told to draft a letter that, said one State Department official, didn't exceed the "bounds of governmental intrusion." Then he added, "But we knew we had to keep the Saudis happy."

The cover letter from Christopher made it clear that the U.S. government was not trying to impose "censorship over the Public Broadcasting System" but requested that PBS give "appropriate consideration to sensitive religious and cultural issues involved and assure that viewers are given a full and balanced presentation." Christopher also enclosed a letter from Alhegelan and a "list of particulars" that enumerated the specific complaints against the film.[20]

The State Department's attempts to influence domestic television programming at the behest of a foreign power were unprecedented—the previous year Turkey had tried to kill the movie *Midnight Express,* but the State Department refused to get involved—and posed a striking contrast to the unwillingness of the British government to intervene. Moreover, as wryly noted by writer Lars-Erik Nelson of the *Daily News:* "In sending the letter to PBS, the Carter Administration, which came into office as an advocate of human rights around the world, found itself defending as 'sensitive religious and cultural issues' a grandfather's right to execute his granddaughter and behead her lover."[21]

At the L'Enfant Plaza office, Chase, having expected much worse, breathed a heavy sigh of relief as he read the long-awaited but relatively restrained letter from Acting Secretary of State Warren Christopher. Grossman had advised him that if the State Department deemed the show not to be in the national interest, he would have to pull the show. Yet there was still one major hurdle to clear. Overseeing PBS decisions is the PTV-1 programming committee made up of PBS station managers and nonbroadcast professionals. It was scheduled to convene that evening, May 8. Although the committee normally would not get involved in such specific programming decisions, the extraordinary outside pressure brought to bear on PBS caused members to make an exception to this long-standing policy.

A conference call hooked up Chase in Washington, Grossman who was out west on PBS business, and the PTV-1 members. The meeting was convened out of Chicago by PBS chairman Newton Minnow, a chairman of the Federal Communications Commission in the 1960s and a leading voice on behalf of First Amendment rights. Minnow was also an early and eloquent critic of typical commercial television programming, labeling it a "vast wasteland." But Minnow was now a partner in the prestigious Chicago law firm of Sidley and Austin—one of the largest in the country.

Within the next year, Sidley and Austin would open offices in Oman, Egypt, and the United Arab Emirates, and would have a partner in its London office attending to Middle East clients and investors. According to one of the participants in the teleconference, Minnow expressed his regret that Mobil had placed its ad in the *New York Times,* but for unexpected reasons: This participant said he recalled Minnow saying that had Mobil not published its ad, PBS could have seriously considered "pulling the show." PTV-1 then affirmed the decision to broadcast the film on May 12.

In order to mollify the Saudis, PBS, under strong pressure from the State Department, agreed to air a sixty-minute postbroadcast discussion featuring former ambassador to Saudi Arabia James Akins ("there are no secrets in Saudi Arabia"); Judith Kipper, a Middle East specialist from the American Enterprise Institute, a conservative Washington think tank supported substantially by oil company contributions; Peter Iseman, a writer considered friendly to Saudi Arabia; Roger Fisher, Professor of

International Law at Harvard University; Ambassador Clovis Maksoud of the League of Arab States; Muddassir Siddigui, Islamic religious leader in Boston; and Andrew Duncan, author of an investigative book on Saudi Arabia. Predictably, none of these "experts," with the exception of Duncan, supported the film; most of the panelists sharply criticized it.

Major stumbling blocks still existed—and would ultimately prevent the broadcast from being shown on all of PBS's 142 local affiliates—for the same tense drama that had been played out in Washington was now being repeated on the local level where numerous public television stations had to contend with enormous pressure against televising the film.

In many ways the power and extent of the petrodollar connection was no more visible, vivid, and ominous than on this level. Not only did local corporate underwriters attempt to kill the broadcast, but in many communities, Mobil's advertisements and the attendant publicity elicited a ground swell of fear.

In western Michigan the Grand Valley State College offices of WGVC-TV were deluged with protests. "The vast majority of the calls have been emotional. They appear based on the fear that, by showing the program, we will cause further deterioration of relations with Saudi Arabia," said Charles R. W. Furman, the station's program manager.[22] The station management logged in 254 calls; only 47 callers supported the decision to broadcast. Many of the objections concerned fears about an oil cutoff.[23] The phone calls were in part triggered by an unusual editorial broadcast numerous times in the days preceding the film by a fundamentalist Christian radio station, WJBL. The station beseeched its listeners to contact the television station— WJBL conveniently provided the telephone number—and ask that the film not be shown: "To purposely offend a country that could bring America to its knees simply by shutting off our oil supply or allowing the price to rise beyond our means is pure folly."[24]

In the end, WGVC-TV withstood the pressure and broadcast the film. So did the Minneapolis public television station, which received more than 500 calls, with a ratio of 2 to 1 against the broadcast.[25] The predominant message of those opposed was the same around the country: Don't jeopardize our gasoline supply. In New York, according to PBS sources, WNET was hit particularly hard by none too discreet warnings from major corporate underwriters. But the station televised the show. However, sixteen other stations did not show *Death of a Princess.*

In Jacksonville, Florida, WJCT pulled the film at the last minute. Why? General Manager Fred Rebman claimed that it had heeded the warnings of "the President [Carter] who warned it was injurious to the lives of Americans."[26] In fact, the President never made any such statement at the time. The real reason WJCT capitulated, according to Jacksonville sources, may have had to do with the presence in Jacksonville of the Charter Oil Company, a giant multinational with sales of $4.3 billion in 1979. Charter had already acquired a reputation for its willingness to do Saudi bidding in the United States. In 1974, Raymond Mason, the head of Charter, devised a scheme that involved a $7.7 million surreptitious public relations campaign, Project Faisal, on behalf of Saudi Arabia. The project called for planting pro-Saudi stories in the Charter-owned publications *Redbook* and *Family Weekly,* producing and showing "documentaries" on college campuses and in theaters, and appearances by pro-Saudi commentators on television shows like "Face the Nation." But Project Faisal came to an unceremonious halt after it was exposed by columnists Jack Anderson and Les Whitten.

Though WJCT denied at the time that Charter had lobbied to kill the film, the station, according to sources, was in the midst of a major fund-raising drive to help pay for a new multimillion-dollar building and studio. And the station had planned to solicit these funds from corporate underwriters, including Charter, which already was a major supporter of WJCT. Said one local television observer who monitored the episode, "this station did not need to be pressured. They just instinctively knew what was in their best interests."

In South Carolina, the five public television stations under the aegis of South Carolina Educational Television (ETV) also dropped the broadcast—a full two weeks before Mobil's ad appeared and the State Department letter became available. On April 24, Charles Morris, director of South Carolina ETV program operations declared that *Princess* would not be shown. "The violence is the main reason" he said in a newspaper interview with the *Columbia Record*. When Peter McGee, director of public affairs programming for Boston station WGBH, which had co-produced the film, was informed of Morris's objection to violence in the film as the basis for canceling it, he responded: "There's more violence in *A Midsummer Night's Dream* than in *Death of a Princess*. This is cowardice."[27]

Two weeks later, Pat Dressler, a spokeswoman for South Carolina ETV, explained that the film had been canceled not only because it was considered "offensive" to the royal family, but also because "the Saudis are people we are relying on for economic resources and we want to be friendly to them."[28] After the decision was announced, the U.S. ambassador to Saudi Arabia, John C. West, previously South Carolina's governor, called public television officials in Columbia to congratulate them for canceling what he found to be a "completely distorted, inaccurate film."[29]

South Carolina firms had landed a particularly large share of the Saudi petrodollar market starting with a colossal $4.5-billion project awarded in 1976 to a joint venture of the California Ralph Parsons Company and Daniel International of Greenville, to build the new Jidda airport, which was to be as large as Manhattan Island. Afterward, scores of other South Carolina firms succeeded in penetrating the Middle East market, thanks in large part to the efforts of John West. By March 1978, West's connections had proved so valuable that reporter Jan Stucker of the *Columbia Record* was moved to comment that Saudi Arabia "has become a popular mecca for Palmetto State entrepreneurs."[30] (See Chapter 18 for more about John C. West.)

Two states to the west, in Alabama, the core of Birmingham's top business establishment waged a vigorous and successful campaign to stop the broadcast.

Princess was blacked out in the nine statewide stations controlled by the Alabama Education Television Commission (AETC). In the week preceding the scheduled broadcast, William Harbert, then executive vice-president of the Harbert Corporation, visited AETC officials in an effort to kill *Princess*. He was concerned about "anything that might upset the equilibrium of relations" between Saudi Arabia and the United States.[31] In 1979 Harbert, one of Alabama's largest multinational construction companies, had signed a $50-million contract to build a water supply system for the city of Jidda. Officials of Blount International, another multinational construction firm with huge contracts in Saudi Arabia, also contacted AETC. Ultimately, Birmingham corporate executives made over forty telephone calls protesting the broadcast. Even a representative of the Birmingham Area Chamber of Commerce joined the effort.

Apprised by officials of Alabama public television of the intensive anti-*Princess* campaign, Jacob Walker, chairman of AETC, hastily arranged an emergency telephone conference at 3:45 P.M. on Friday afternoon, May 9, with four of the five commission members. Though none had seen the film, the members eventually made a tentative decision not to show it. (One commission member, however, argued unsuccessfully in favor of the film's broadcast.) On Saturday morning, May 10, AETC announced—after Walker conferred with John Harbert, chief executive officer of the Harbert Corporation—that the film would not be aired, attributing this decision to the rather novel concern for the safety of Alabama citizens in Saudi Arabia: "The film was so sensitive that it might endanger the well-being of several hundred Alabamians working in Saudi Arabia and other Arab countries."[32] Even Alabama Congressman William Dickinson, whose district includes Montgomery where Blount is located, became involved, contacting PBS in Washington to try to stop the national broadcast.

In Texas, where the Port of Houston registered a whopping 2.4 billion dollars' worth of exports to the Middle East in 1979, petrodollar trade had filtered down to the actual owners and licensees of public television stations—the universities. Consequently, the stations needed little if any prodding from the business community not to show the film. One station, based in College Station, had determined to show the film but was overruled at the last moment by its licensee. At 2:45 P.M. on May 12, just hours before the scheduled national broadcast, the management of KAMU received a startling order from its licensee, Texas A&M University: *Princess* would not be shown. An accompanying statement delivered by Texas A&M University President, Dr. Jarvis Miller, stated:

> *The decision to forgo running this particular program was made on the basis that we should not risk damaging international relations by showing a movie that reportedly relies on sensationalism and shock value to attack a culture and religion that is foreign to us. As a university we are attempting at this very time to establish significant new ties with the people who are most offended by this movie.''* [33]

One hundred miles to the southwest, at Station KUHT, funded and operated by the University of Houston, the situation was pretty much the same, except that the decision not to broadcast *Princess* had been made at least twelve days before the national broadcast date. The station's programming director and general manager had favored showing the film, but he was overruled in an unprecedented decision made by Patrick A. Nicholson, vice-president for public information and university relations of the University of Houston. This was the first such programming decision made by Nicholson in his seventeen-year tenure at the university.

Nicholson issued a press release attributing the decision to the "strong and understandable objections by the government of Saudi Arabia at a time when the mounting crisis in the Middle East, our long friendship with the Saudi government and U.S. national interests all point to the need to avoid exacerbating the situation." When Nicholson later testified in an unsuccessful suit brought against the University of Houston by a Houston resident to compel the broadcast of *Princess,* the court found that Nicholson was "entirely unable . . . to explain what he meant by this phrase ['exacerbating the situation']." There were other reasons why Nicholson canceled the show—reasons he neglected to include in his press release. Several years earlier, the university had signed a lucrative contract with the Saudi Arabian royal family to provide full-time instructors for a princess in Riyadh. And, in addition, having been responsible for the university's fund-raising activities between 1957 and 1978, Nicholson had become acutely aware that "a significant percentage of the university's private contributions came from major oil companies and from individuals in oil-related firms." [34]

Ironically, Saudi objections contributed to making *Death of a Princess* one of the most popular PBS programs of all time. In New York the film attracted 26.8 percent of the television audience, making it the most popular program in its time period, beating out favorites such as "Little House on the Prairie," "Flamingo Road," and "That's Incredible." In Chicago the film placed second in the ratings.

Yet Saudi Arabia did have the last word. Aside from the one-hour post-broadcast discussion—which, in the words of one senior PBS official, consisted of "Sixty minutes of Arab apologia"—PBS immediately broadcast a special program that painted a highly favorable portrait of the status and role of women in Saudi society. The effort to appease Saudi anger succeeded. [35] An internal PBS memorandum in June 1980 noted that Mrs. Faisal Alhegelan, the wife of the Saudi ambassador, agreed to appear on a special PBS program "with the approval of her husband." Broadcast on June 5, 1980, the program examined the role of women in Saudi Arabia from a perspective that was intentionally different from that of *Princess.* Moreover, State Department consultations immediately began with the Saudi government for a major documentary series on Saudi society and foreign policy. Two years later, the highly flattering one-sided series was broadcast across the United States and has been repeated several times. In contrast, *Death of a Princess* has never been shown again on PBS.

For the first time since the 1973 oil embargo, a "litmus test" of American-Saudi relations had been put simultaneously to the American executive branch, Congress, and the public. Though *Princess* was ultimately broadcast, the Saudis and their American agents had cause for much celebration. The executive branch and the public had shown themselves susceptible to friendly and coercive persuasion. The jury was still out on Congress.

10

AWACs:
PRELUDE
TO THE
BATTLE

[T]he strength of the Zionist influence
is nothing but a wooden house that can
be broken when America comes first.

Al-Jazira, *a Saudi newspaper, the
day after the U.S. Senate voted
to sell AWACs to Saudi Arabia*

Precisely at 5:03 P.M. on October 28, 1981, the Senate roll call began. It was the moment of reckoning that Washington's political intelligentsia had anxiously awaited. The upper Senate galleries were packed with hundreds of spectators, frenzied journalists, and nervous diplomats from the Middle East. Outside the galleries were thousands more—many who had come to witness those precious few moments of history being made in the nation's capital and many who had participated in some way in the historic AWACs debate. As the name of each senator was called by the president pro tempore of the Senate, their votes were eagerly recorded by the hordes of lobbyists, foreign dignitaries, administration officials, and Senate staff members sitting in the galleries or standing in the anteroom off the Senate floor.

As the third and final roll call began, the tension on and off the Senate floor produced an awed silence. One of the most intense, protracted and bitter political debates since the Vietnam War was about to come to an end. The shock waves, whichever way the vote turned out, would be felt around the world. For on that crisp, dry, historic Wednesday afternoon, the United States

Senate was to decide whether the United States would sell AWACs—the most advanced airborne radar technology system in the world—to Saudi Arabia.

Not only was the vote regarded as a referendum on the power and prestige of the presidency, but also at stake, proponents contended, was America's relationship with the entire Arab world. Yet opponents of the sale were unable to counter with equally significant considerations: the threat to Israel's security; the unacceptability of a presidential fait accompli, especially since the Reagan administration did not, as it promised, allow Congress to participate in proposing the sale or in determining its configuration; the fears of many senators (reminiscent of the Iran debacle in which the superior Phoenix missile technology was lost to the Russians in the political turmoil after the shah fell from power) concerning the transfer of such sophisticated technology to an autocratic and potentially unstable country; and the destabilization of the delicate balance of power in the Middle East.

The AWACs—Advanced Warning and Command systems—are the most advanced command-and-control radar systems in the world, generations ahead of the best Soviet technology. The AWACs serve as a combination early warning system, battlefield surveillance, and tactical battle control system. Each AWAC can track and direct 240 fighters at once; separate friendly from hostile aircraft, and provide their range, bearing, altitude, speed, and other data; compute fifteen interception solutions every second; track friendly and hostile ground troops, missile sites, airfields, and even ammunition storage centers; make "threat assessments" and suggest the highest-priority targets; relay data to friendly planes and to anti-aircraft missile and cannon ground stations; and detect up to six hundred targets. Displayed instantaneously to AWACs technicians, the data can be relayed via secure communications to friendly jet fighters and interceptors, ground defenses, and command stations, where all incoming data can be integrated.

Further enhancing the awesome capacity of AWACs is its "down-linked" communications system. The "look-down" capability allows the radar to separate the images of the aircraft from the radar echoes off the ground, thus enabling the radar to "see" downward and to differentiate low-altitude moving targets (low-flying planes, for instance, or even cars) from terrain or "background clutter." The down-linked communications system renders communications immune to hostile electronic jamming. AWACs flown in tandem with the F-15 air superiority fighter—a plane already provided to the Saudis by the United States—equipped with its own "look-down" radar provide devastating military capability. The AWACs–F-15 combination forms the cornerstone of the NATO tactical air defense against the Soviet Union. Even the AWACs guarding Europe are maintained exclusively under U.S. control.

But now the Senate was being asked to confer upon Saudi Arabia the extraordinary privilege of ownership and control over the AWACs. The bitter nine-month debate, which commenced in the very first days of the Reagan administration, was finally coming to an end. Congressional sentiment against the sale had been overwhelming from the moment the decision was announced in March 1981. The balance of forces arrayed against the sale remained stable until two days before the vote. Then suddenly the opposition started to crumble.

In a few moments it would be all over. Yet the excruciating situation in which many senators found themselves was painfully similar—and needlessly so—to the wrenching F-15 debate they had waged three years earlier. The outcome of that debate was supposed to have ended the spiral of escalating Saudi "litmus tests" of American friendship toward the kingdom. But the subsequent political peregrinations of the Carter administration had ironically ensured that, in the words of one senator who prefers to remain unidentified, "Jimmy Carter's chickens would come home to roost on Ronald Reagan's doorstep." The behind-the-scenes machinations of the Carter administration also revealed beginnings of the spectacular Saudi leverage over executive branch policy, leverage that would continue to grow despite the change of administrations in 1980.

In March 1978 President Jimmy Carter, in an abrupt change in American foreign policy, agreed to sell to Saudi Arabia sixty advanced F-15 Eagle air superiority aircraft. Congress, however, balked at the President's proposal. Realizing that the sale faced certain death in the Senate, the President agreed to modify the terms of the sale and provide assurances that the weapons could not be used against Israel in a larger Arab-Israeli dispute in which Saudi Arabia would be drawn in.[1] And so, in assuaging congressional concern, the Carter administration provided Congress in May 1978 with explicit guarantees that the F-15 craft would be used only for defensive purposes and would not be followed by the sale of offensive equipment. Specifically, the Carter White House promised Congress that it would not sell auxiliary fuel tanks that could expand the range of the F-15; the devastating AIM 9-L all-aspect Sidewinder missile, which unlike other air missiles, could hit enemy aircraft from all positions rather than seeking the rear end heat-emitting side; the omnipotent radar surveillance plane, AWACs; and "any other systems or armaments that could increase the range or enhance the ground attack capability of the F-15."[2] In justifying their affirmative votes, over twenty senators invoked the President's assurances that the sale would be defensive only and that it would not be followed by any future sales of offensive ordnance. In addition, the President had tied the sale of the F-15s to the sale of jet fighters to Israel and Egypt. Congress was thus faced with an all-or-nothing vote. If the F-15s to Saudi Arabia were defeated so too would be F-16s to Israel and F-5s to Egypt. The sale of the F-15s to Saudi Arabia and the accompanying planes to Israel and Egypt was approved by a vote of 54–44.

Less than two years later, however, in January 1980, the Saudis, according to officials in the State Department, specifically requested the right to purchase those weapons that had been denied them and that they had agreed not to seek. Their request was received sympathetically by White House officials. In a visit to Riyadh in January 1980, National Security Adviser Zbigniew Brzezinski reportedly told the Saudis that their request for arms would be met sympathetically by the Carter White House. In the following three months, consultations between the Defense Department, the State Department, and Saudi officials were conducted in order to prepare for submission to Congress a package of new Saudi arms. This package was to include fuel tanks; multiple ejection racks, which permit the F-15 fighter-bombers to carry larger loads of bombs; KC-135 tankers for refueling the F-15s in midair; the advanced air-to-air AIM-9-L Sidewinder missile; and AWACs. Though aware of likely congressional resistance to the sale, especially in the light of its own commitments to Congress, Carter administration officials professed their inability to say no to the Saudis. One high-ranking unidentified White House official, quoted in the *New York Times,* said, "They've thrown down the gauntlet. We need a lot of things from the Saudis, and there is not too much we can do but grant the request." [3]

Despite a flurry of newspaper reports mentioning or citing the proposed sale, the Carter administration refused to consult Congress or answer repeated requests from members of the Senate Foreign Relations Committee to confirm or deny the newspaper reports. [4] Fearing a presidential fait accompli, in which Congress would be foiled again with an all-or-nothing Saudi arms package, members of the Senate sent President Carter a letter on July 8, 1980, reminding him of the commitments his administration had made two years before. The letter effectively killed any chance that the sale would go through. Reacting to the Senate letter, "a senior member of the Saudi ruling family" attributed congressional opposition to the "Zionist lobby" in Washington. He added, "we believe that with a little more courage and foresight all American administrations . . . could have stood to the Zionists' pressures." [5]

Yet the Carter administration was not to be deterred. In fact, under strong Saudi pressure the administration succeeded in

watering down the text of the letter as it was being drafted before the final version was distributed to members of the Senate
for signature.[6] Even before the November elections had taken
place, the Carter administration set in motion the institutional
process to sell offensive equipment to Saudi Arabia. Defense
Secretary Harold Brown ordered a formal comprehensive "review" of the Saudi request for additional equipment. Not surprisingly, the outcome of the perfunctory Defense Department
"review" was one of strong support for the Saudi position. Now
the only question was whether President Carter would act affir-
matively on the Defense Department recommendation to approve the vast Saudi military shopping list.

The President decided in favor of the Saudis. In several
meetings that occurred in December of 1980 between outgoing
President Carter and President-elect Reagan, Carter offered to
take the political heat and grant Saudi Arabia the right to purchase the new weapons. In this way, according to Reagan officials who were privy to the conversations between the two men,
Reagan could profess to be angry with Carter, but could look
the other way. However, declining to accept Carter's offer,
Reagan told the departing President that his administration would
evaluate Middle East policy, including arms for Saudi Arabia,
from scratch.

The new Republican administration was considered sympathetic to Israel. Specifically, many of its new officials were
inclined to view Israel as an asset, not only because of its stability but also because of its strategic position as a buffer against
Soviet expansion in the Middle East. Moreover, given the new
administration's ideological prism, which viewed almost all regional conflicts as flowing from an East-West rivalry, Israel's
security was not to be tampered with. Finally, perhaps the most
important reason why it was thought that the new administration
would not shift the military balance of power away from Israel
was that Ronald Reagan had been an avowed friend of Israel
ever since its creation in 1948. During the 1967 Six Day War,
Reagan helped raise funds for Israel at an Israeli Bonds rally in

California. And in the 1980 campaign, candidate Reagan specifically criticized President Carter for waffling in 1980 on the 1978 commitments regarding offensive F-15 equipment.

So it came as a complete surprise to many observers when the new Secretary of Defense Caspar Weinberger said in his maiden news conference on February 4, 1981, that "we want to make them [the F-15s] as effective as we can for their [the Saudis'] purposes." Weinberger strongly suggested that the Reagan administration would approve the sale of offensive F-15 equipment to Saudi Arabia.

Even though Weinberger was suspected of having an Arab "bias" due to his previous employment with Bechtel, a firm with billions of dollars' worth of contracts in Saudi Arabia, the feeling in Washington was that Weinberger was not venting official policy, only his own. But, during the following week, the administration leaked a bombshell: in its very first foreign policy initiative, the administration had decided to sell to Saudi Arabia the offensive F-15 equipment. The deal, however, was not to include advanced AWACs. Yet, as news of the package leaked to the press, eight members of the Senate Foreign Relations Committee sent a letter to President Reagan on February 20 expressing their "serious concerns about the possible sale of such equipment." On February 26, Under Secretary of State James L. Buckley provided informal briefings to the House Foreign Affairs and Senate Foreign Relations committees on the new arms proposals. In order to allay Israeli fears, Buckley reported that Israel would be compensated with the opportunity to purchase fifteen additional F-15s. For its part, Saudi Arabia let it be known to officials in the administration that it regarded the sale of the enhanced equipment as "a litmus test" of its relations with the United States.

Congressional opposition to the sale emerged from the entire political spectrum. Undaunted, however, the Reagan administration formally announced on March 6 that it intended to sell Sidewinder air-to-air missiles and extra fuel tanks to Saudi Arabia. At the same time, Israel would be given an additional $600 million in military credits to help compensate for its increased vulnerability. Although Israeli military analysts protested that Israel would be compelled to establish a new line of air defense if Saudi Arabia obtained the advanced F-15 equipment, the government of Israel decided that it could not fight the sale, and quietly let it be known that it would not attempt to rally its supporters in Congress. Yet, despite Israel's acquiescence, opposition to the sale continued to grow, some of it coming from senators not known for their sympathy for Israel. Day after day, groups of senators would deliver colloquies against the sale. On March 24, for example, a bipartisan group of twenty senators spoke out repeatedly against the sale of the offensive ordnance—particularly in light, they said, of Saudi Arabia's attempts to undermine the Camp David Accords. Administration officials were surprised at the vehemence of congressional opposition, especially since the Israeli government had decided not to oppose the sale. By April 22, more than forty-eight senators had spoken out against the arms package.

Meanwhile the National Security Council (NSC) held a meeting on April 2 to discuss the sale of weapons to Saudi Arabia. It had been expected that the NSC, in light of the congressional opposition, would arrive at a compromise. Instead, the council, which was chaired by Vice-President George Bush in the absence of President Reagan, who was recuperating from gunshot wounds, agreed to provide Saudi Arabia not only the offensive F-15s but the five AWACs as well. The projected date for submission of the package to Congress was April 27. As word of this decision leaked out, Congress went into an uproar. On April 7 over a hundred Republican and Democratic congressmen blasted the proposed sale in two hours of speeches on the House floor. The opposition to the AWACs came not only from senators concerned with the security of Israel but also from those who feared that the sophisticated AWACs technology would be compromised if given to Saudi Arabia. Assistant Republican Senate Leader Ted Stevens, an ally of the President, candidly declared that the sale to Saudi Arabia was "in real trouble."

Reagan's best friend in the Senate, Nevada Senator Paul Laxalt, conceded, "I think at this point a majority of the Senate would probably be opposed to the sale to the Saudis of the whole package, including the AWACs."

Opposition to the sale continued to snowball. According to an Associated Press poll released on April 20, senators opposing or leaning against the sale to Saudi Arabia outnumbered those in favor by more than two to one. Sixty-five senators said they were either committed to vote against the sale or were leaning that way. AP could find only twenty senators who would vote in favor of the sale or would be predisposed to consider it. It was clear to the President and his aides that the sale was in deep trouble. The administration's plan to submit the arms package to Congress on April 27 had to be postponed. So, on April 26, Republican Majority Leader Howard Baker announced that President Reagan agreed to delay sending the package of equipment to Congress. Baker predicted that a vote would take place in the fall.

Congressional leaders, however, suspicious that the administration would not seek concessions from the Saudis and thus present the Senate with a fait accompli, continued to voice their concerns. On June 25, Senator Bob Packwood, the maverick Republican from Oregon who was spearheading the opposition to the sale, announced that fifty-four senators had signed a letter to the President opposing "the Administration's intent to sell the F-15 enhancement and AWACs package to Saudi Arabia. It is our strong belief that this sale is not in the best interests of the United States, and therefore recommend that you refrain from sending this proposal to the Congress." That same day Congressmen Clarence Long (Democrat of Maryland) and Norman Lent (Republican from New York) released a House resolution disapproving of the sale; the resolution was signed by 224 members of the House.

It was clear to administration pulse-takers that the opposition to the sale was widespread and deep. And it was also clear that the opposition came from many lawmakers simply concerned with the possibility that the AWACs technology would not be safe in Saudi Arabia. In spite of these sobering conclusions, the administration intended to forge ahead. In a private and unannounced meeting held with Prince Bandar bin Sultan, the son of the Saudi defense minister, President Reagan promised that the AWACs package would be submitted to Congress in the early fall.

Yet there could be no denying that the weapons proposal was in deep trouble on Capitol Hill and the President's advisers knew that the sale would not pass. Indeed, the administration had all but given up trying to win over the House of Representatives. Resigned to insurmountable opposition in the House, White House officials focused their attention on the Senate. Under the terms of the Arms Export Control Act, once Congress receives formal notification of an arms sale, it has thirty days to stop that sale by a majority vote in both houses. If they split, the sale goes through.[7] So on September 14, the President commenced his personal efforts in behalf of the sale by inviting twenty-seven senators to the White House. The reaction, however, was far from encouraging. Many of the senators, especially the Republican freshmen, expressed their firm opposition to the sale. In its September 28 issue, *Newsweek* reported that fifty senators and half a dozen "signatories in absentia" signed a bipartisan petition against the AWACs sale.

Realizing the difficulties at hand, the administration tried several times during the month of September to prevail upon Saudi Arabia to compromise. At one point, Secretary of State Alexander Haig announced that he intended to travel to Saudi Arabia to seek concessions from the Saudis in the operation of the planes. But the Saudis canceled that trip abruptly and firmly told the administration not to delay submitting the package to Congress. Said one official close to Saudi Arabia to me, "the Saudis said they would go to other sources such as England and France to buy what they needed if the administration postponed the sale. After all, what does a country with a fourth of the world's oil reserves need to say after they have said that?" Pressed by Saudi

Arabia, the administration informed Congress that it would submit a formal notification by October 1.

In testimony before the relevant Senate committees, Secretary of State Haig and Secretary of Defense Weinberger encountered stiff resistance. In one meeting of the Armed Services Committee, Senator Dan Quayle, a young Republican freshman from Indiana who had been swept to victory in the 1980 Reagan landslide, bluntly told Caspar Weinberger that the AWACs sale had no chance of Senate approval: "What do we need to do to convince the administration that this package is not going to go through as it currently is?" Determined to force a showdown with Congress, however, President Reagan formally notified Congress on September 27 of his administration's intention to sell the $8.5-billion package of sophisticated radar planes and other offensive equipment to Saudi Arabia.

Within three hours after the proposal was submitted on October 1, Senator Bob Packwood submitted a resolution of disapproval with the signatures of fifty senators, one short of the majority needed to block the sale in the Senate. *Time* magazine reported that "nose counts by both sides" showed only twelve senators in support of the sale, with sixty-five against.[8] On October 14 the House voted overwhelmingly, 301–111, against the sale. Both the House Foreign Affairs Committee and the Senate Foreign Relations Committee voted to disapprove the sale. But on October 28, less than one month after the President had submitted the sale, one of the most stunning upsets in post–World War II congressional history occurred. Defying all predictions, the proposal to sell AWACs and offensive F-15 equipment to Saudi Arabia squeaked through the Senate by a 52–48 vote. Only two days before, the *New York Times* had reported the Senate lineup as 53 against the sale and 38 in favor.[9]

The willingness of so many senators to change their minds astounded even their colleagues. Senator Edward Kennedy, for example, who had voted against the sale, spoke for many observers when he commented, "In my 19 years up here I have never seen such 180 degree turns on the part of so many Senators. I have seen nothing that would justify this kind of reversal."[10]

Later that evening, the victors celebrated at a party hosted by the Tunisian ambassador at his residence. Over twenty-five ambassadors from Arab, African, and European nations attended the joyous reception along with State Department officials, corporate executives, and senators. Special toasts were made in honor of Texas Senator John Tower, chairman of the Armed Services Committee, and Illinois Senator Charles Percy, chairman of the Senate Foreign Relations Committee, both of whom had helped shepherd the AWACs through their committees and through the Senate at large.

Back in Saudi Arabia, officials were jubilant over their victory. The lead editorial in the Saudi newspaper, *Al-Jazira,* declared, "[T]he strength of the Zionist influence is nothing but a wooden house that can be broken when America comes first."

11

THE AWACs VOTE: FARMERS, FLORISTS, AND RICE GROWERS

The AWACs sale will provide a general spur
to the economy and it will also put some
more people to work in this country. And
the more people working in this country
and the more they have in their pockets,
the more they will spend on flowers.

Spokesman for a florist
insurance company

In early September 1981 a group of five well-dressed men met regularly but inconspicuously in a plush eight-room suite at the Fairfax Hotel (now the Ritz Carlton) in downtown Washington. Without any fanfare, they at first met three mornings each week, but as September advanced, the frequency increased to five mornings a week and some evenings as well. Though their presence may not have seemed out of the ordinary, the same could not be said of their mission.

These five men were the principal figures involved in charting a course that would send shock waves around the world for many years to come: these men were plotting out the core of the lobbying strategy needed to secure congressional approval for the sale of the $8.5-billion AWACs to Saudi Arabia. Their central purpose was to elicit the aggressive involvement of the entire spectrum of the American business community. It was the most ambitious lobbying goal ever pursued by American business on a foreign policy issue.

Located on fashionable Massachusetts Avenue, just west of Dupont Circle, the Fairfax Hotel, a favorite hangout for Saudi royalty, is known for its elegance and its cosmopolitan atmosphere. Its well-heeled guests include high-powered business-men, Hollywood luminaries, and government leaders from around the world. A dinner for two at less than $125 is considered a real bargain.

The group consisted of three principal registered agents for Saudi Arabia—Frederick Dutton, Stephen N. Conner, and J. Crawford Cook—and two Saudis—Prince Bandar bin Sultan, son of Saudi Defense Minister Sultan, and Abdulah Dabbagh, a former commercial attaché at the Saudi embassy in Washington. The Saudi ambassador never sat in on these daily meetings, but he would often arrive just before the group met in order to discuss matters with Prince Bandar. Though Dabbagh is not a member of the royal family and no longer worked for the Saudi government, his extensive business contacts in the American corporate community made him a critical player in devising the AWACs game plan.[1]

Prince Bandar, who officially became the Saudi defense attaché in 1982 and has since become the Saudi ambassador, was brought over to the United States in early 1981 in anticipation of the political battle over AWACs and F-15 equipment. Good-looking, debonair, and extremely charismatic, Bandar instantly became a star of Washington's elite cocktail circuit, entertaining and being entertained by numerous senators, journalists, and key administration officials. Secretary of State George Shultz, for example, has been a tennis partner of the young Saudi prince. Bandar has also been a racquetball partner of the head of the Joint Chiefs of Staff, General David Jones. A neighbor of Senator Edward Kennedy, Bandar lives in a $1.6-million mansion in McLean, Virginia. The importance of this young Saudi can best be gleaned from a statement included in a *Wall Street Journal* profile of him: "[W]hen Prince Bandar calls for a Central Intelligence briefing the analyst comes to him."[2]

For almost nine months Reagan administration officials had been reassuring the Saudis both in Washington and in the kingdom that eventual passage would be no problem. Yet the opposition to the AWACs was more entrenched than ever. The participants at the Fairfax Hotel meeting realized early in September that the Reagan administration could not be relied on to secure passage of the AWACs in Congress. The Saudis and their American agents had firmly decided that if the job was to be done, they would have to do it themselves.

Part of the strategy was to create an impression that Bandar, other Saudi officials, and the American agents were unobtrusive presences, that they had limited political influence and were around only to answer questions, while allowing the media to focus disproportionately on the role of the Jewish lobbying.[3] The *Wall Street Journal* wrote, "Prince Bandar's role, however, is primarily that of messenger and good will agent. He hasn't taken a leading role in lobbying Senators." The *Journal* quoted Bandar: "I'm here to be available for any senator who wants to know our side of the story . . . but we are being low key. We aren't fighting this battle, the President is."[4] Bandar actually played a very active role. He took the initiative in making the rounds on Capitol Hill, setting up meetings and arranging briefings with forty senators. Between September 8 and September 12, for example, he met with John Tower, Charles Percy, Barry Goldwater, Richard Lugar, Dale Bumpers, Harry Byrd, and S. I. Hayakawa. He developed a close working relationship with Senate Majority Leader Howard Baker, flying down to Baker's guest house in Huntsville, Tennessee, in early September for a tête-à-tête with Baker and Under Secretary of State James L. Buckley. Baker also provided Bandar with a regular office off the Senate floor.[5]

In interviews with the press, Dutton downplayed the lobbying efforts of the Saudis: "You don't see Arab lobby groups going up and down the corridors of Congress. Why? Because they can't rely on a nationwide constituency, nor can they provoke a flood of telegrams."[6]

Dutton also attempted to portray the national debate over AWACs as one that pitted the interests of Israel against those of the United States. "If I had my way," he told the *Washington Post*, "I'd have bumper stickers plastered all over town that say

'Reagan or Begin,' " in an attempt, the *Washington Post* wrote, "to exploit what he [Dutton] perceives as widespread resentment of the idea that a foreign chief of state could exercise a kind of veto over Presidential decision-making."[7] And, in an interview with me after the AWACs vote, Dutton said, "They didn't do the basic lobbying. They didn't want people like me to go to the Hill [Congress], which, for any lawyer in Washington sounds nonsensical."

In its October 12 issue, *Time* magazine, which three weeks later intentionally ignored a major story on Saudi-influenced lobbying, played down the efforts of the Saudis compared to the activities of the American Jews:

> *The American Israel Public Affairs Committee has been presenting a detailed case against the sale to Senators and Congressmen for the past five months. The lobbying also has a strong grassroots component, with Jewish leaders in each state who agree with the lobby's aim being asked to contact their Congressman. For their part, the Saudis have hired a high-priced Washington consultant, Frederick Dutton, but efforts on their behalf have generally been quiet.*[8]

In fact, the three Saudi agents, who were paid over $1 million for their services in 1981, worked furiously behind the scenes to marshal support for the AWACs sale. At the direction of Saudi Arabia, the agents served as the principal contact point between Saudi officials and officials of the White House, National Security Council, State Department, Defense Department, and Congress. A special working team from the administration met regularly with all three agents. Senator Baker and his staff met with the Saudi officials and their agents on over fifty separate occasions, mostly in downtown Washington restaurants or on Capitol Hill, during the four months preceding the vote.

The agents also attempted to generate or orchestrate support from other quarters. Among the organizations whose representatives met with Cook—whose offices were located in the Madison Hotel opposite the *Washington Post*—were the Washington-based Center for United States–European–Middle East Cooperation, Americans for Middle East Understanding, National Association of Arab Americans, and the Vietnam Veterans Organization.

Conner arranged for a private meeting on September 19, in Palm Springs, California, between Prince Bandar and former President Jerry Ford in which U.S.-Saudi relations and the AWACs sale were discussed. A month later, Ford telephoned various senators and expressed his support of the AWACs sale.

Three weeks later, on October 5, Ford, as director of the Santa Fe International Corporation, voted in a unanimous board of directors' decision to ratify the Sante Fe president's invitation to Kuwait to purchase the giant $2.5-billion oil and drilling company. As the new owner of Sante Fe, Kuwait gained instant access to Sante Fe's subsidiary, C. F. Braun and Company, which developed nuclear technology. Despite the clear potential dangers attending such a sale—the Congressional Office of Technology Assessment declared that C. F. Braun's technology "would be of use to any nation that wishes to produce nuclear weapons"—Ford approved of the Kuwaiti purchase. The price was double the stock value.

Both Cook and Conner helped arrange a pro–AWACs sale news conference on September 22 at the Hyatt-Regency Hotel on Capitol Hill for former ambassadors to Saudi Arabia, John West, Parker Hart, James Akins, and Robert Neumann. Portrayed to attending journalists as spontaneous political action by former high-ranking U.S. officials, the press conference had been set in motion four days before at a private luncheon at the Fairfax attended by the four former ambassadors and Prince Bandar.

Most important, the three Saudi agents realized that if the AWACs sale was to pass the Senate, the mobilization of the entire American business community would be essential. At the Fairfax meetings, lists of corporations and chief executive officers were broken down by state and matched with the names of senators considered politically vulnerable. Saudi contracts with American firms were compiled as were lists collected of all American companies doing business in Saudi Arabia. Senator Baker provided the latest head counts in the Senate. And political intelligence on the progress of the lobbying efforts in Congress and in the business community was exchanged with other centers of corporate organizing: the Business Roundtable, United Technologies Corporation, Westinghouse, the Boeing Company, NL Industries, Mobil Oil, Bechtel Corporation, and the American Businessmen's Group of Riyadh.

But Cook and Conner, in the words of one informed source, "worked the hell out of the private sector. They were the hot and heavy guys." Hundreds of chief executive officers and corporate presidents and vice-presidents were contacted and strongly urged to write or call their senators. Prince Bandar and other Saudi officials made daily phone calls to chief executive officers and also met with many in Washington, asking them to help by talking to their senators. Meetings were also held in Paris, London, and other European cities where Saudi officials, sometimes accompanied by the registered agents, would inform visiting American corporate executives of the importance placed on the AWACs sale by the Saudi government.

In the six weeks preceding the Senate vote, scores of Washington representatives of major American corporations were invited to attend "receptions" at the Saudi embassy and other Saudi-designated locations where the importance of the sale to Saudi Arabia was communicated along with the reminder of the companies' extensive business ties in Saudi Arabia.

Many American corporations were led to believe that much of America's trade with the Saudis, including their own, would disappear if the sale did not go through. After conversations with Prince Bandar, other Saudi officials, and the registered agents, some corporate chiefs were left with the unambiguous impression that their lobbying efforts would actually be monitored. Some businessmen were pointedly told that the success of their efforts might decide their company's future involvement in Saudi Arabia. In other instances the promise of contracts was dangled in front of corporations, as one corporate official said, "like raw meat before a hungry dog."

By October, however, prospects for passage of the AWACs, despite administration pressure, dimmed rather than improved as everyone had expected. Indeed it seemed that the chances for passage of the AWACs sale in Congress were worse than ever. So a formal decision was made to increase the pressure. In an unprecedented move, Saudi Arabia, according to officials close to the kingdom, held up final contract negotiations with American firms during the time the sale was before Congress.[9] All contracts awaiting final signatures were frozen, and discussions of new contracts were postponed. Several companies were warned that if the sale did not go through or if they did not lobby hard enough, renewals of their contracts would be in jeopardy. With one exception, no American contracts in Saudi Arabia were awarded or renewed during the period from October 5 to October 28, 1981.

The one exception was the Whittaker Corporation. In August 1981 Whittaker's president, Joseph Alibrandi, published an article titled "If I were a Saudi Arabian" in *Newsweek*'s "My Turn" column.[10] Alibrandi extolled the altruistic and pro-Western policies of Saudi Arabia, praised Saudi Arabia as "the primary peacemaker and defender of American interests in the Middle East," described how Saudi Arabia has purchased American technology and invested in American securities as a favor to the United States, and attacked the West's ungrateful response. Under his byline, Alibrandi was identified very simply as a "corporate president, [who] has done business in Saudi Arabia for nearly twenty years."

Missing from his brief biographical description were more pertinent details of Alibrandi's background. From 1960 to 1970, Alibrandi was an executive with the Raytheon Company stationed in Saudi Arabia where he helped install ground-to-air Hawk missiles for the Saudis. When he took over Whittaker in 1970, it was a diversified manufacturing company with almost no experience in hospital management. But Alibrandi's contacts in the Saudi Ministry of Defense and Aviation soon helped his new company enter the health care field in a significant way. In 1974 the Whittaker Corporation was awarded a $100-million contract to manage and operate three Saudi hospitals, and the contract eventually grew to $500 million. By 1981, the company employed over 2,000 people in Saudi Arabia.

After the *Newsweek* article appeared, Alibrandi mailed a copy to every member of the U.S. Senate. On October 26, 1981, two days before the scheduled vote on the AWACs sale, Whittaker announced that Saudi Arabia had awarded it a contract to expand the corporation's health care program to include five hospitals. The new fee was $834 million, a 67 percent increase over its existing contracts. Whittaker's Saudi business provided 46 percent of the company's pre-tax profits for 1981. Several months later Whittaker became embroiled in a bitter fight over its attempted takeover of the Brunswick Corporation. It offered $329 million in cash, but Brunswick resisted. In a newspaper advertisement, Brunswick charged that "Whittaker is subject to the influence of Saudi interests. Sheikh al-Fassi of Saudi Arabia has been quoted (in the *Washington Post*) as saying, 'this is one of the biggest companies in the world and I control it. Joe Alibrandi is my employee.' "[11]

Whittaker was not the only corporation to receive a lucrative contract after actively promoting the AWACs sale. In mid-October, the Arizona-based Greyhound Corporation contacted the state's two senators to inform them that the company regarded the AWACs sale as important. Greyhound became involved, according to an official, because "the AWACs controversy had the potential to become disruptive for all, economically and politically." The official added that "it is not our style to knuckle under to any type of pressure" and that "it was a matter of conscience. The sale of AWACs would serve the country, the U.S. public, and this corporation as well."[12]

Eight days after the sale of AWACs was approved by the Senate, Greyhound Food Management, a wholly owned subsidiary of the Greyhound Corporation, announced that it had been awarded a $90-million contract extension by the Saudi Ministry of Defense and Aviation. Greyhound officials later insisted that the contract extension was formally signed in late September, but that the press release went out in November because the mail "does not move quickly between the United States and Saudi Arabia." (This excuse was rather unconvincing in an age of instant communications and in light of the fact that Greyhound actually operated an office in Saudi Arabia.)

On July 18, 1981, an official of National Medical Enterprises, a health care management corporation based in California, wrote to all members of the California delegation to Congress and to some other out-of-state members. "My perception," wrote the official, "is that the sale is in the best interests of the United States." In September 1981, Saudi Arabia signed an $84-million contract with National Medical Enterprises for management of a hospital. A company spokesman later explained that the lobbying was the "personal work" of one official based in Saudi Arabia who "unfortunately wrote on company letterhead."

Three weeks after the AWACs vote, the Westinghouse Corporation announced that it had been awarded a contract for construction of a power plant in Hail, Saudi Arabia, worth $130 million. Five Westinghouse spokesmen in Washington and the corporate headquarters based in Pittsburgh vehemently and repeatedly denied to me that any lobbying had occurred. In fact, during the month of October, Westinghouse lobbyists provided Senate head counts to the National Association of Arab Americans, while its Washington representatives conferred with Saudi agent Conner. Moreover, Westinghouse—which made the computers for the AWACs—took the unusual step of retaining the public relations firm of South and Baroff for a fee of $75,000, according to a Westinghouse memorandum, "to conduct public opinion research and to consult on media matters" in helping to garner support for the AWACs.

As the lobbying effort gathered steam, there was greater

coordination among the corporate participants. Representatives of over forty companies, primarily in areas related to defense, aerospace, and petroleum, attended meetings held in the downtown Washington offices of the Business Roundtable, the country's most powerful business lobby. The meetings were organized by Richard M. Hunt, director of government relations for NL Industries, a manufacturer of petroleum equipment and supplies. Hunt denied that NL Industries was motivated by any desire for economic gain, telling me, "we appreciated the role Saudi Arabia played in maintaining moderate oil prices and also the need to maintain Middle East stability." [13]

In the beginning of September, Hunt's coalition met once a week. But by the beginning of October, the meetings were being held much more frequently, sometimes as often as four times a week. Hunt refused to discuss details of the meetings, but according to one corporate representative, the sessions became intense and frank. The White House sent a staff person from its business liaison office who, like the organizers of the meetings at the Fairfax Hotel, determined which senators were considered vulnerable to political pressure. Toward the end, questions of ethnicity were mentioned explicitly—like which senators were Jewish. And in one of the final sessions, according to an executive who attended the meetings, a representative of a major corporation rose before his hushed Business Roundtable colleagues and said, "the children of Israel will stub their toes on this one."

Apart from Hunt's leadership role in organizing the ad hoc coalition, NL Industries executives met separately with four senators and contacted thirty-nine others by letter and Mailgram (from states where NL Industries had facilities) to press them to support the AWACs sale. A year and a half later, in its first quarter 1983 report mailed to stockholders, the company—responding to a stockholder resolution—candidly admitted that business interests dictated its involvement: "The company has operations in Saudi Arabia and other Middle Eastern countries and therefore has an interest in the peace and prosperity of the region. A reversal of the previously announced U.S. position on AWACs could have imperiled that objective.''

Two corporate giants that became heavily embroiled in the lobbying campaign were Boeing, the main contractor for the AWACs plane, and United Technologies, which estimated that it had some $100 million at stake. According to their spokesmen, Boeing President E. H. Boullion and United Technologies Chairman Harry Gray together sent out more than 6,500 telegrams to subsidiaries, vendors, subcontractors, suppliers, and distributors throughout the country. Boullion's telegram asserted that ''without this sale and with present budget planning the AWACs production line will be ended.'' Boullion also added that ''a negative decision on this issue may offset Saudi Arabia's attitude toward other products.''

Gray, in addition to mailing thousands of his own telegrams ''requesting'' support for the sale, pressured the chiefs of his subsidiaries—including Pratt & Whitney, Carrier Air Conditioning, Otis Elevator, and Ideal Electric—to send out the same telegrams to hundreds of their vendors and distributors. The result was one of the most successful chain-letter operations in history. One vendor said that the head of a United Technologies subsidiary told him confidentially that in the end over ten thousand and possibly as many as twenty thousand telegrams were sent to businesses across the country. ''Anyone who had ever done business with United Technologies was asked to support the sale,'' he added. Ultimately, letters filtered down to valve companies, small businesses with fewer than twenty employees, and even some mom-and-pop industrial distributor operations.

In his telegram, Gray couched his support for the AWACs sale in lofty terms:

> The Saudis are a force for stability and moderation in the Middle East. . . . A Senate veto of the air defense package would cost our nation dearly in a weakened relationship with a staunch friend in the Arab world, diminished Presidential influence for

world peace and stability, lower U.S. credibility in
the world's capitals, and lost U.S. exports and jobs.

Although Gray appealed for support on foreign policy grounds, his vice-president, Clark MacGregor (formerly chairman of Richard Nixon's reelection committee), appealed to Republican fears of losing their newly gained Senate majority. Three years before, MacGregor had worked discreetly with Stephen Conner to help pass the sale of the F-15s, but now his role was much more visible. To Republican leaders and Republican opponents of the AWACs sale MacGregor sent last-minute warnings that the Republicans would lose their majority in the Senate in the 1982 elections if the sale was defeated: "Should President Reagan's plan be disapproved next week by the Senate—with the decisive votes being cast by Republicans—the results would be devastating to the 1983 hopes of maintaining a Republican majority in the Senate."

Though MacGregor's telegram was typical of the scare tactics frequently employed in high-pressure lobbying campaigns, the same could not be said for Harry Gray's telegram. It ended on an omnious note: "Please wire your two U.S. Senators today asking them to sustain the President's position. Would you also send me a copy of your communication to the Senators?" Corporate officials, many dependent on United Technologies' goodwill, responded overwhelmingly. I reviewed more than two thousand letters sent in by corporate supporters of the AWACs sale. More than fourteen hundred of these telegrams and letters paraphrased or quoted the points made in the United Technologies and Boeing telegrams.

Some firms copied the telegrams verbatim; others tried to enhance their authenticity by claiming in an introductory paragraph that the impetus for writing came out of "internal discussions" with all the employees. One corporation had every one of its twenty officers send in the same telegram. Other officials included this line at the bottom of their telegram: "cc: Harry Gray, United Technologies Corporation." To many senators, who

received weekly mail counts, the number of corporate supporters of the AWACs surely seemed impressive. But two companies, Boeing and United Technologies, may have been responsible for generating approximately 70 percent of the "grass roots"
corporate mail.

The pressure exerted on the recipients of the United Technologies' telegrams was staggering. One vendor, with annual sales
of under $15 million, said that he was subjected to "raw eco-
nomic blackmail." He continued, "at our level all small vendors are engaged in an intense economic competition with each
other. Very few have a big edge over the other in price—the
final selection is sometimes arbitrary. So when Harry Gray asks
vendors to lobby on the AWACs, what choice do we have? How
do we know that our refusal to support the AWACs won't be
used against us later?"

A year and a half after the AWACs vote, the sixty-four-
year-old United Technologies executive talked unabashedly about
his desire to gain additional Saudi contracts. The occasion was
a lavish Washington reception at the Textile Museum, where a
Saudi exhibition, "A Caravan through Arabia," was on display. United Technologies picked up the $25,000 tab for the
evening. Said Gray, who attended the reception and was pho-
tographed next to the wife of the Saudi ambassador, about his
company's sponsorship of the event, "We have some projects
in Saudi Arabia. We'd like to get more." [14]

The lobbying campaign also drew on the resources of
construction, engineering and oil companies with large Saudi interests. Three prominent firms in this area were Brown and Root;
Bechtel; and Aramco. Brown and Root, for example, which has
worked on Middle East projects totaling more than $100 millon,
supplied pro-AWACs position papers to at least a dozen Sunbelt
senators—from Texas, Oklahoma, Louisiana, Alabama, Mississippi, and Arkansas.

Bechtel, the world's largest construction and engineering
firm, derived nine percent of its 1980 gross income from its Saudi
projects, constituting its largest foreign market, according to

Department of Defense documents. Besides contacting senators directly—Bechtel letters referred to the AWACs sale's enhancement of the "U.S. national security interests"—Bechtel also played a critical behind-the-scenes role in eliciting the involvement of subcontractors, vendors, and suppliers, which were dependent on Bechtel's goodwill. However, in his confirmation hearings some eight months after the AWACs sale was passed, Secretary of State–designate George Shultz, president of Bechtel, scoffed at the statement made by Senator Larry Pressler, Republican of South Dakota, that Bechtel had engaged in intense lobbying. "We let our position be known," but that, Shultz claimed, was the extent of Bechtel's efforts. He categorically denied having pressured any affiliated business to lobby.[15]

Another major source of corporate pressure and organizing activity was the American Businessmen's Group of Riyadh, headed by an official of the Northrop Corporation stationed in Saudi Arabia; members of this organization had contributed a half-million dollars to Senate candidates in 1980. The group jumped to an early start, having sent out a nine-page memorandum to its members on June 7, 1981, promoting the AWACs sale. The memorandum stated: "Each of you knows how important your business in Saudi Arabia is to your company. This message must be communicated to the Senators and Congressmen from the states and the districts of your firms and your vendors at all levels."

The American Businessmen's Group of Riyadh is an amalgam of the top *Fortune* 500 companies, such as Citibank, American Express, Great Atlantic and Pacific Tea Company (A&P), and Merrill Lynch International. To ensure that Congress was acutely aware of the membership, a list identifying the 123 member companies was published and prominently distributed on Capitol Hill by the National Association of Arab Americans. The six largest senatorial recipients in 1980 of the $500,000 given to Senate candidates by the American Businessmen's Group of Riyadh all supported the sale.

The John Deere Company, manufacturer and distributor of

farm equipment and machinery, targeted twenty-three senators, mostly from the Midwest, who received letters and/or phone calls from the top executive officers. Senator Charles Grassley, Republican of Iowa, for example, was blitzed with letters from the president and chairman of the board of John Deere and followup letters from other high-level executives of the company. And in a letter sent to Charles Percy, chairman of the Senate Foreign Relations Committee, a Deere vice-president wrote, "In recent years Israel has become increasingly disruptive and I support President Reagan's serving notice that their continuing interference in U.S. foreign policy is not welcome." (Percy voted for the AWACs sale.)

W. H. Krome George, chairman of Alcoa (Aluminum) wired all U.S. senators in support of the AWACs on "behalf of the employees of Alcoa." Robert Dee, chairman of the Smith-Kline Corporation, a health care company known principally for its manufacture of pharmaceutical products, personally wrote every member of the U.S. Senate urging them to vote for the sale of the AWACs. William Delancey, chairman and chief executive officer of Republic Steel, contacted twenty-six senators who represented states in which Republic has facilities.

In lining up support in Congress, Northrop's president, Thomas V. Jones, met with Prince Bandar on September 15. The Saudi agents conferred separately with a steady stream of corporate officials including the head of the governmental affairs unit for J. C. Penney Company, one of the nation's largest clothing and general merchandise retailers, and representatives of the General Electric Company, Fairchild Industries, FMC Corporation, and Northrop.

The Halliburton Company, a Texas-based energy firm, contacted a dozen Sunbelt senators. Dresser Industries, a petroleum equipment and energy technology producer based in Houston, also lobbied southwestern senators.

The pro-AWACs lobbying campaign drew upon the resources of the entire spectrum of the American business community—those with and without interests in Saudi Arabia. The list of those that lobbied comprises a veritable Who's Who of major executives and corporations throughout the United States. These include, to pick at random, Dravo Corporation (construction and engineering); Centex Corporation (a multi-industry holding company); Overseas Advisory Association (consultants for the U.S.–Saudi Arabian Joint Commission); Barnes & Tucker (coal mining); Owens–Corning Fiberglass; Fisher-Price Toys Incorporated; Cooper Industries (manufacturer of compressor, drilling, and electric products); Poppe Tyson, Incorporated (industrial advertising); Cubic Corporation (farecard machines and other high-technology products); American Standard, Incorporated; PVI Industries, Incorporated (a small business manufacturer of commercial water heaters); and even some Pepsi-Cola bottling companies.

One of the oddest recruits to the AWACs cause was the Association of Wall & Ceiling Industries International, a nonprofit association representing American contractors from carpenters to plasterers. The group's interest in radar planes for Saudi Arabia would appear to be somewhat tangential. Nevertheless, its executive vice-president, Joe Baker actively promoted the sale of AWACs within the business community. Baker denied that he got involved in the AWACs campaign in any significant way other than to "communicate with a single senator asking him to vote for the AWACs approval." He said, "I did this on my own as an American citizen and one who strongly supports President Reagan. The Association has never taken a position for or against the sale of AWACs to the Saudis." [16] But Baker's letter to Senator Grassley was written on official stationery from the Association of Wall & Ceiling Industries. Moreover, a White House staff member and Richard Hunt of NL Industries both singled out Baker for his active support for the AWACs sale. According to White House sources, and confirmed by Baker, President Reagan even wrote him a letter thanking him for his help. In an internal memorandum circulated to the executive committee of the Association of Wall & Ceiling Industries, Baker told of his "active" involvement in the "AWACs matter," though he attributed his activity to his "support [for] the President." [17]

Nor was the petrodollar connection limited to oil, manufacturing, retail, and construction industries. Powerful sectors of the agricultural community—including the rice lobby, grain companies, and farm co-ops—also became aggressive supporters of the AWACs sale. Riceland Foods, a major rice producer and distributor based in Arkansas, lobbied Arkansas Senators David Pryor and Dale Bumpers, as well as local congressmen, during their visits back home in September and October 1981. Officials of American Rice, Incorporated, P&S Rice Mills, and other rice growers based in Texas, contacted members of the Texas delegation to Congress.

In perhaps the most extraordinary move among the rice lobby, the Rice Millers Association, the rice industry's trade association, held a board of directors meeting on September 18, 1981, and voted to support the sale of AWACs to Saudi Arabia. It then contacted senators from Florida, Mississippi, Missouri, Arkansas, Louisiana, and Texas.

Why this unprecedented involvement by rice growers in the sale of weapons? A Riceland Foods official said it became involved because it was concerned with "foreign policy, Middle East peace, and trade." A spokesman for the Rice Millers Association initially cited the need for "stability in the Middle East" and said that "peace could be established by the sale of the AWACs." He then added that the sale of the AWACs was "just a matter of economic sense." American rice exports to Saudi Arabia from August 1980 to July 1981 totaled $150 million, he said, and represented a 50 percent increase over the previous year. The sale of rice to Saudi Arabia was "strictly a cash market," and "the Saudis paid premium prices."

Cenex, the eleventh largest farm cooperative in the United States, lent support to a lobbying effort that included retaining a Washington law firm to promote the AWACs sale on Capitol Hill. With 1981 sales of $1.3 billion at the time, Cenex was part of a consortium with four oil refiners known as Interdependent Crude and Refining (ICR).[18] ICR delivered a slick brochure, prepared by its managing director, to all senators. The managing director, John Venners, also visited the offices of at least a dozen senators in support of the AWACs sale. Why? According to the brochure distributed by ICR the consortium was moved to act because

> the Saudis are asking for our help—a display of our faith and support, a renewal of trust, and a commitment to friendship . . . even though we do not purchase oil from Saudi Arabia, we are concerned. We're concerned because we are all well aware of the strategic importance of Saudi oil . . . concerned because of the unsettled nature of the region and the threat to attack. . . . We cannot afford to lose their [Saudi] friendship and continued cooperation.

There were other reasons, too. ICR was formed in mid-1981 for the express purpose of obtaining a direct crude oil supply contract with Saudi Arabia. But by September 1981, ICR had not met with any success. According to sources, Venners told members of the consortium that if ICR could tangibly demonstrate its friendliness to Saudi Arabia, the Saudis would respond with a contract. So the consortium decided to go all out in support of the AWACs sale. According to a person familiar with ICR, "it would provide a great entrée with the Saudis if they could prove that they were in the forefront on the lobbying efforts. Up until then, ICR couldn't even get the Saudis to look at them [ICR]."

Cenex's involvement is even more interesting in light of the fact that Cenex is a member of the National Farmers Union, a progressive general farm organization. Many of the National Farm Union's board members were in fact opposed to the AWACs sale. But Cenex approved ICR's decision to lobby in support of the sale anyway. According to the *Political/Finance Lobbying Reporter*, an authoritative Washington newsletter, a spokesman for Cenex also said, "We're not opposed to lobbying . . . our members need the crude." [19]

According to Senate sources, packs of officials representing farm co-ops—many of which have substantial investments in petroleum products, refineries, gasoline stations, and fertilizer production—roamed the Senate halls looking for senators opposed to the sale of AWACs. Several senators and their staffs were warned, according to Senate sources, "if you want sixty thousand angry letters tomorrow from co-ops, then keep up your vote against the AWACs!"

Even farmer organizations became involved. The National Farmers Organization (NFO) is a collective bargaining organization designed to secure the highest prices for farmers and producers. NFO's Washington director, Charles Frazer, sent out twenty letters on NFO stationery to senators expressing support for the AWACs sale. Although the board of NFO never took a position, Frazer reportedly had been told to proceed from key individuals in NFO. Another high-ranking official in the farm community from the Midwest said that "discussions over whether to support the AWACs were held throughout the Midwest farm co-op community. AWACs was a big issue for us."

Frazer of the NFO was not the only one to move without the formal blessing of the organization for which he worked. Richard Lesher, president of the U.S. Chamber of Commerce, sent out letters to all senators the day before the vote, in which he reported on his October trip to the Middle East. Lesher noted in his letter that the sale of AWACs was "not a one-nation, isolated issue. Everywhere we traveled to discuss improved trade and foreign relations, the AWACs issue always emerged—always calling for support." He wrote that "recommendations for approval" of the AWACs sale came from American businessmen and foreign officials in all the countries he visited, which included Saudi Arabia, Morrocco, United Arab Emirates, Kuwait, and Egypt.

When I asked about Lesher's lobbying for the U.S. Chamber of Commerce, a spokesman explained, "Across the board there was unanimous consent that if the sale did not go through, there would be negative repercussions." Though the U.S. Chamber of Commerce never formally took a position, it did hold a reception in October for Saudi Minister of Commerce Soliman Sulaim, during which Sulaim asked business to support the sale. In addition, the chamber included in its 860,000-circulation newsletter sent out in September an article that warned of the adverse consequences for U.S. trade if the sale was defeated.

Even the Florists Insurance Companies sent letters in support of the AWACs sale to various senators and representatives. "The AWACs sale will provide a general spur to the economy and it will also put some more people to work in this country," a spokesman explained to me. "And the more people working in this country and the more they have in their pockets, the more they will spend on flowers." According to this spokesman, letters of opinion are routinely sent out by the Florists Insurance Companies to congressmen on various issues, but acknowledged that this was its first active venture into foreign policy.

In late October, with the Senate vote still hanging precariously in the balance, a traveling contingent of American corporate elite decided to provide added push for the AWACs sale. A group of top corporate executives toured Eastern Europe and the Persian Gulf on a "fact-finding mission" sponsored by Time, Inc., similar to the one sponsored by Time in January 1975 (see Chapter 3). Accompanying the traveling executives were sixteen *Time* magazine editors, correspondents, and company officers. The group landed in Riyadh on October 27. On the morning of October 28—the day of the AWACs sale vote—a confidential telex was sent from Riyadh to a select group of twenty senators. The telex, signed by twenty-three of the twenty-four visiting executives, pleaded for approval of the sale claiming that

> *a negative vote on the AWACs sale would:* (1) *severely damage U.S. credibility in the Arab world;* (2) *substantially diminish U.S. ability to play a critical role in assisting peace negotiations between Israel and*

Arab interests; (3) permit the Soviets to increase its influence in the Arab Gulf and support its argument that the U.S. is an unreliable ally; (4) substantially impair U.S. ability to protect its legitimate interests in the Middle East in the future. . . . It is our opinion that rejection of the sale would be a great disservice to the American people, to U.S. prestige and to the cause of peace in the world.

Many corporations whose officials signed the telex have done business in Saudi Arabia, some have offices located there, and nearly all are hoping to expand their markets in that country. When asked about the telex, one corporate spokesman suggested it was Time's idea. In fact, it has been learned that Theodore Brophy, chief executive officer of General Telephone & Electronics (GTE), came up with the idea of sending a telex on the night of October 27. According to a participant on the trip, Brophy discussed his proposal with a few colleagues, and asked a correspondent whether Time would be put in an embarrassing position if the Time-hosted corporation group sent letters "to pressure their senators" to support the AWACs sale.

The next morning the travelers were given a briefing by an American official in the U.S. embassy in Saudi Arabia. This provided an occasion for all of the travelers to be together. At that point, Brophy stood up and suggested to the group that they send a telex. As Brophy drafted the text of the telex, Henry Grunwald and other *Time* correspondents left the room. One of the corporate participants wondered about the possible embarrassment to the group if the telex were disclosed publicly, especially since it emanated from Riyadh: "Wouldn't it appear suspicious given where we are right now?" But this objection was immediately dismissed, and with the exception of one noncorporate leader, Vernon Jordan, the telex was signed by all the junketeers.

Early in the afternoon the group was treated to a lavish tour of the palace of Prince Abdullah ibn Abdul Aziz, commander of

THE RIYADH 23

The names and titles of the corporate officials who signed the telex from Riyadh to senators on the morning of the AWACs vote:

Robert Anderson
Chairman, Chief Executive Officer
Rockwell International Corporation

John R. Beckett
Chairman of the Board
Transamerica Corporation

Theodore F. Brophy
Chairman, Chief Executive Officer
General Telephone & Electronics Corporation

Philip Caldwell
Chairman, Chief Executive Officer
Ford Motor Company

Albert V. Casey
Chairman, Chief Executive Officer
American Airlines, Inc.

Richard P. Cooley
Chairman, Chief Executive Officer
Wells Fargo Bank, N.A.

Donald W. Davis
Chairman, Chief Executive Officer
The Stanley Works

Edwin D. Dodd
Chairman, Chief Executive Officer
Owens-Illinois, Inc.

Myron DuBain
Chairman, President, Chief Executive Officer
Fireman's Fund Insurance Companies

Henry J. Heinz II
Chairman of the Board
H. J. Heinz Company

Matina S. Horner

T. Lawrence Jones
President, Chief Executive Officer
American Insurance Association

Robert E. Kirby
Chairman, Chief Executive Officer
Westinghouse Electric Corporation

William E. LaMothe
Chairman, Chief Executive Officer
Kellogg Company

William S. Litwin
President
KeroSun, Inc.

Stewart G. Long
Vice President–Sales & Services
Trans World Airlines, Inc.

Henry Luce III
President
The Henry Luce Foundation, Inc.

Robert H. Malott
Chairman, Chief Executive Office
FMC Corporation

John J. Nevin
Chairman, President, Chief Executive Officer
Firestone Tire & Rubber Company

Paul C. Sheeline
Chairman, Chief Executive Officer
Intercontinental Hotels Corporation

John G. Smale
President, Chief Executive Officer
The Procter & Gamble Company

Thomas J. Watson Jr.
Chairman Emeritus
International Business Machines Corporation

L. Stanton Williams
Chairman, Chief Executive Officer
PPG Industries, Inc.

the 30,000-man Saudi national guard. At that particular moment, however, since Crown Prince Fahd was outside the country and King Khalid was sick, Abdullah, as the highest-ranking member of the royal family, was in control of the kingdom. Abdullah lectured the group about the role played by Jews in the United States. In an acerbic tone, Abdullah, obviously posturing, said he hoped for a defeat of the AWACs sale: "I am personally hoping for the failure of the vote today. It would make them realize that there is another government [Israel] that influences American policy." Abdullah also told the group that "American aid to Israel" constituted the greatest threat to Saudi Arabian security.[20] Two hours later, at 5:00 P.M., after the group left the palace, some of the travelers again asked that the senators from their home states be contacted once more. Heinz Company chairman Henry J. Heinz revealed to the group that his son, the senator from Pennsylvania, would vote for the AWACs sale if his vote was needed.

Brophy's leadership in organizing the AWACs lobby evoked curiosity from one of the participants. Significantly, until 1980 GTE was subjected to the Arab boycott. Some political observers believe that Brophy may have been trying to demonstrate to the Saudis the correctness of the Arab decision to drop GTE from the boycott list.

Though the signers wrote that their appeal was "personal" and that they were responding "individually," each of them (except Matina Horner, who signed the appeal but did not identify herself as president of Radcliffe College and a member of the board of Time, Inc.) provided his title and the name of his corporation.

In an interview later given to the *Harvard Crimson* after her lobbying role was disclosed, Matina Horner said that she saw "no danger" in the sale of AWACs because they are "clearly a defensive piece of equipment." You can't do anything offensive with an AWACs."[21] Horner added that she would have been "much more concerned had the United States contemplated selling nuclear equipment." In a separate letter responding to criticism of her decision to lobby for the sale, Horner gave a different reason: "I did, however, sign the petition on the basis of *high level confidential information* which encouraged me to believe that under the circumstances, a positive vote on the AWACs sale would be in the best long-term interests of the United States and the peace in the Middle East"[22] (emphasis added). Horner has never revealed this "high level confidential information."

Thomas Watson, chairman emeritus of IBM, was much more blunt about the need to avoid alienating Saudi Arabia. In a letter he sent out to a critic of his lobbying, Watson wrote, "It was apparent to me that if the deal was not approved we would so alienate Arabians at all levels that our future negotiations with them would be very difficult."[23]

The sale was approved by a 52–48 vote, although at least four other senators had promised to vote for the sale if their votes were deemed necessary. The corporate lobby was decisive in achieving that margin, reversing the overwhelming opposition to the sale that had existed only one week before the vote.

One midwestern senator had been called by every chief executive officer in his state. Utah Senator Orrin Hatch, a one-time leader in the anti-AWACs coalition who ended up voting for the sale was hit very hard by corporate leaders, including Harry Gray of United Technologies, who personally lobbied him.

One key Senate staffer who was on the winning side on the AWACs vote said that Senators Christopher Dodd of Connecticut, Lowell Weicker of Connecticut, Alfonse D'Amato of New York, Charles Grassley of Iowa, Roger Jepsen of Iowa, Mack Mattingly of Georgia, and Thad Cochran of Mississippi were all hit particularly hard by corporate interests, especially by United Technologies, whose political action committee was Connecticut's wealthiest in 1980 ($153,180). Of those seven senators, four voted for the AWACs.

In Nebraska powerful corporations generated enormous pressure on Senators Edward Zorinsky and James Exon, both Democrats. On the day of the vote, Zorinsky conceded, "I've got everyone who's got a vested interest economically both in the State of Nebraska and nationally who's got bucks to make or lose in the event that the sale is turned down.[24] Valmont Industries, Leo A. Daly, and Henningson Durham and Richardson—firms with offices and/or business in the Middle East—lobbied the two Nebraska Democrats, as did Conagra (an agricultural and commodities firm), Internorth, and First National Bank—firms without substantial interests in the Middle East. The chairman and chief executive officer of Valmont Industries, Robert Daugherty, helped organize the lobbying campaign in Nebraska and was congratulated by a "top Saudi official" the morning after the vote.[25]

Zorinsky had been a co-sponsor of the resolution against the sale, and Exon was reported to have been leaning solidly against the sale, but both succumbed to the fierce pressure. Exon was pressured by the president of Union Pacific Railroad and by Omaha banking executives. In explaining more bluntly what had influenced his vote, Zorinsky told a Washington lobbyist, in the confines of his Senate offices, that he had discovered that the lampposts in the city of Riyadh were made in Nebraska.

Howell Heflin, Democrat of Alabama, was confronted by a delegation of twenty-seven Alabama businessmen, many with contracts in Saudi Arabia, such as Blount Incorporated, which had been awarded the largest fixed-price construction contract ($1.7 billion) in history in 1981. Some of these businessmen met with the registered Saudi agents to rehearse their speeches before their meeting with Heflin. According to one source, the message Heflin received was clear: "Either you support the AWACs or you are our ex-senator from Alabama." In the end, however, Heflin voted against the AWACs.

Roger Jepsen, Republican of Iowa, had been a staunch leader in the anti-AWACs coalition, but he was turned around at the last moment under pressure from farm manufacturing equipment companies, farm co-ops, and defense contractors. In 1982 he collected $11,000 in honoraria from speeches to major defense contractors that supported the AWACs sale. These firms included Pratt & Whitney, Raytheon, McDonnell Douglas, and Martin Marietta Corporation.

One major company, the FMC Corporation, whose CEO, Robert H. Malott had been one of the earliest Saudi proponents in the United States, initiated a full court press. A midwestern senator said that the president of FMC "took my head off" at a party in Chicago when the FMC executive discovered that the senator was set to vote against the sale of AWACs. Another FMC official—its vice-president—sent letters to twenty senators who had FMC facilities in their states. Still other FMC executives initiated appointments with four senators and fifteen Senate staff members to make their position known in person. At one meeting with Senator Grassley, Republican of Iowa, who remained opposed to the sale until the day of the vote, an FMC representative emphatically informed him that his company considered the sale of AWACs "its number one foreign policy issue."

David Boren, Democrat of Oklahoma, had been solidly leaning against the sale, but in the end, he voted for it. He and Senator John Melcher of Montana had been subjected to enormous pressure from oil interests in their states. According to the *Wall Street Journal,* oil men told Senator Boren that the United States could not afford to offend its major supplier of oil.[26]

In 600 letters that I examined (excluding the 1,400 letters and telegrams that originated from the massive chain-letter operation organized by United Technologies and Boeing), I found over 150 that expressed, in varying degrees, anger at American Jews and their lobby groups. One letter, for example, from the chief executive officer in the Cubic Corporation, Walter Zable, who, incidentally, had been induced into lobbying by the Boeing Corporation, wrote all U.S. senators: "It doesn't take much study to see the necessity for supporting Saudi Arabia, and Congress should not take too seriously the words of lobbyists and organizations which distort the picture. And why should Israel continue to dictate to the U.S. on the course of action we should take?" A president of a medium-sized construction firm wrote,

"Israel's small geographic size and Jewish population, estimated at only 2 million, as well as its general lack of natural
resources, leave only its pugnacious capabilities for military action and a strong U.S. lobby as serious considerations [against
the sale of AWACs].'' And a vice-president of a major international investment bank complained in a telegram to at least
one senator: "I'm tired of having Jews run our foreign policy.
And you should be too."

The Mobil Oil Corporation deserves special mention, both
as one of the prime corporate propagandists for the Saudis and
for its possible violation of the Foreign Agents Registration Act.[27]
Spending over a half-million dollars, Mobil produced a media
blitz in early October, publishing a series of full-page advertisements in twenty-six major newspapers and magazines across the
United States.

What was most interesting about these advertisements was
the conspicuous absence of any reference to the AWACs sale or
even arms sales in general. Rather, Mobil focused exclusively
on the "profound and rapidly growing economic partnership between the United States and Saudi Arabia." Mobil presented this
message in fine detail, informing its readers, for example, that
Saudi business extends to forty-two states, that "hundreds of
thousands" of jobs have been created by Saudi contracts, that
$35 billion will accrue to the U.S. balance of trade as a result
of Saudi friendship and that more than seven hundred companies are now doing business in Saudi Arabia. Mobil even asserted that the Saudis' willingness to invest a "major part of their
surplus funds in dollars" stemmed more from political favorit-
ism toward the United States than from economic self-interest.
In its final line Mobil summed up its message very neatly: "Saudi
Arabia is far more than oil—it means trade for America, jobs
for Americans and strength for the dollar." In another full-page
advertisement, Mobil simply listed the names of two hundred
U.S. corporations involved in Saudi Arabia under the headline
"$35 Billion in Business for U.S. Firms."

In addition to avoiding any mention of the AWACs sale,

the advertisements also represented a sharp departure from previously published political "commentaries" on the Arab-Israeli conflict. In fact, in late August 1981, Mobil printed a column in its "Observer" format, praising selected statements by Yasser Arafat—even misrepresenting Arafat's words—and heaping more praise on the Saudis for their "peace plan." In the last line, Mobil called for "even-handed treatment in the sale of military equipment to our friends in the Middle East, especially Saudi Arabia."

In shifting its advertising focus from an "analysis" of Middle East politics to the economic advantages of Saudi trade, Mobil hit upon a successful theme for its brazen campaign to engender American goodwill for the Saudis. As one official close to the Saudis and their lobbying efforts said to me, "Mobil did more for the AWACs through their ads than anyone else." In light of the record showing that previous pro-Saudi public relations activities were initiated at Saudi request, the most recent media campaign raised legal questions about whose interests Mobil was representing. When asked by a *London Times* reporter whether the Saudis had applied pressure to the company to lobby, a Mobil official said, "Overtly, none." Then he changed this to, "No, none at all." [28]

Most corporations that lobbied on the AWACs attempted to keep their involvement secret, except, of course, from the two groups of people that counted the most—members of Congress and the Saudis. One sourse close to the Saudis said that even if the sale had not gone through, the U.S. corporations could have shown the Saudis evidence of their loyalty: "It was a process of building up chits."

Many corporate officials denied taking any position on the sale, but when presented with hard evidence of their involvement, they tried to dismiss it as a spontaneous expression of civic responsibility. Upon further questioning, however, especially when asked if the Saudis communicated with them, they would immediately fall back to a second line of defense and readily—

almost too readily—concede that their activities were spurred by economic self-interest. For example, when asked whether the Saudis had pressed United Technologies into lobbying, a spokesman for United Technologies unwittingly discredited the entire foreign policy cover extended by its president by telling me, "They [the Saudis] didn't have to. It was a matter of pure economic self-interest."

Most tellingly, it has been discovered that a handful of the major corporations agonized over the issue of lobbying for the AWACs sale and finally decided not to do so. Some of them actually attended meetings of the Business Roundtable and receptions at the Saudi embassy to give the impression that they backed the sale. Spokesmen for these firms have pleaded that they not be identified. One corporate official said, "If our absence on lobbying becomes conspicuous, there will be hell to pay."

From discussions with scores of corporate officials and spokesmen, it is also clear that most were unfamiliar with the substantive issues underlying the controversy behind the AWACs, nor did many of them care. But once these executives were made to feel that their contracts might be in jeopardy, or once they received a call from Prince Bandar, or a request from the registered agents, or the telegrams from Boeing and United Technologies, or the memo from the American Businessmen's Group of Riyadh, or the message from Mobil Oil's editorial ads—they quickly called Western Union and joined the crowd.

Beyond the act of lobbying, the AWACs episode stimulated the corporate leaders' willingness to use their financial clout in Congress to promote their views on the Middle East. In fact, in the aftermath of the AWACs campaign, some corporations began to punish their senators. According to Republican party and congressional sources, major Pittsburgh-based corporations withdrew their participation in a fund-raising event for Senator Heinz out of spite for the senator's vote against the AWACs sale. Another East Coast Republican Senator, William Roth of Delaware, also found that corporations based in his state bought fewer tickets for an annual fund-raising event because of Roth's position against the AWACs sale.

The success of the Saudis, their registered agents, and their chief corporate supporters in eliciting the massive business lobbying represents a watershed in the relationship between the Saudis and their American business partners. The AWACs sale had produced the most extensive business lobbying on any foreign policy issue since World War II. From banks to bus companies and from florists to farmers, they came out in droves. It had been only eight years since some of these Americans angrily denounced the Saudis for their oil embargo and had demanded a reciprocal food embargo. But petrodollar wealth and trade had finally been diffused throughout the United States.

Influential friends of Saudi Arabia had suddenly appeared in abundance and were willing to do almost anything to preserve the special Saudi-American relationship.

12

THE
STATE
DEPARTMENT
HOAX

The Shia minority does not pose a serious
political threat to the regime.

*July 1981 State Department
briefing memo for use before
Congress during AWACs debate*

The Shia minority could pose a major security
threat over the next two to five years.

*December 1980 classified
State Department report
on Saudi stability*

$\mathbf{A}$t 3:01 P.M. on July 9, 1981, the seventh-floor State Department office of James Buckley logged in a priority briefing memorandum prepared by State Department experts. The concisely written memo contained vital background materials for use by Buckley, the under secretary of state for security affairs, in briefing Congress, the press, and the public on why Saudi Arabia was considered such a stable country.

At that particular moment, the sale of AWACs to Saudi Arabia was in deep trouble on Capitol Hill. Opponents of the sale were clearly making headway with the argument that the advanced AWACs technology was far too precious to U.S. national security to be entrusted to a potentially unstable regime like that in Saudi Arabia. Three months later, with the AWACs still being debated, advance copies of an extensive report touting the internal stability of Saudi Arabia were circulated to key congressional committees. The report had been prepared under the auspices of the highly prestigious and influential Washington think tank, the Georgetown Center for Strategic and International Studies, known by its acronym CSIS.

There was one detail, however, that the authors of the State Department memorandum and the think tank report neglected to point out: both reports contained distortions, fabrications, and disinformation. Moreover, the doctored reports had been written and distributed in a calculated effort to deceive the Congress, the press, and the public. The reports had been selectively lifted from an earlier classified report that revealed the instability of the Saudi government. Amid a hotly contested debate that focused extensively on the wisdom of transferring to another country such sophisticated military technology, the likelihood of congressional approval might have been greatly diminished, if not destroyed, had evidence of Saudi instability been brought to the attention of the U.S. senators and the American public.

The classified State Department report was written in December 1980 by one of the foremost State Department experts on Saudi Arabia, David Long, Director of the Office of Analysis for Near East and South Asia in the Bureau of Intelligence and Research. In vivid detail, it revealed that Saudi Arabia was plagued by widespread royal family corruption, religious turmoil, severe problems of loyalty in the military, and the likelihood of ''a major security threat'' by the Shia minority ''within two to five years.''

Through interviews with government officials and from documents obtained under the Freedom of Information Act, I have been able to piece together the details surrounding this political ruse. What has emerged is a story of wrongdoing, fraud, and corruption committed by Arabists at high levels of the U.S. government working in concert with one former government official, exploiting the availability of corporate petrodollar-linked funding. Their scheme resulted in the deliberate dissemination of false information to Congress and the public, abuse of the laws governing declassification, and violation of the integrity of supposedly disinterested research by an academic institution. Senior State Department officials participated in and sanctioned both the preparation of the deliberately inaccurate memo and the selective replication of the classified report. Revelation of the details of this plot provide a case study of how petrodollar influence has corrupted the public policy process.[1]

The roots of this tale date back to 1977. Amos A. Jordan, Jr., then executive director for international resources programs of CSIS (he is now CSIS president), was tracking issues for possible CSIS projects. Saudi Arabia, as an emerging political and economic giant, he thought, deserved closer attention. "I have long followed events in the Middle East," recalled the fifty-four-year-old Jordan, "specifically in Saudi Arabia, and I had a perception that this was too important a subject area—developments in the [Persian] Gulf, specifically Saudi Arabia—for us not to be having some work going on." Jordan has considerable background in and familiarity with the Middle East. He served as principal deputy assistant secretary of defense in 1974, and deputy secretary of state and acting under secretary for security assistance in 1976. In 1977, he joined CSIS.

Founded in 1962 with a seven-member staff, CSIS now has 150 employees, a $6-million yearly budget, and a prestigious position in the Washington shadow-government establishment. Linked by title only to Georgetown University, much of the funding for CSIS is provided by corporations and corporate foundations such as Atlantic Richfield, A. W. Mellon, and the Rockefeller Foundation. CSIS was once home to ex–National Security Adviser Richard Allen, and other officials of the Reagan administration. The neo-conservative think tank currently houses James Schlesinger, Zbigniew Brzezinski, and Henry Kissinger (and his seven-room office suite and twelve-person staff). On Middle East issues, the institution has no set disposition, retaining numerous analysts whose views run the entire political spectrum. During the AWACs debate, however, two CSIS fellows, ex–U.S. Ambassador to Saudi Arabia Robert Neumann and Alvin Cottrell testified in favor of the sale.

When Jordan was at the State Department he befriended John Shaw who became inspector general of foreign assistance and later assistant secretary of state overseeing foreign aid and military sales. As the newly elected Carter administration assumed office, Jordan left the State Department to join CSIS. Shaw stayed on at State for a while, but in 1977 departed to work for Booz Allen and Hamilton International, a leading management consulting firm. The firm offered Shaw a position as head of the thirty-member planning group in Saudi Arabia that was to advise the kingdom on the construction of the two new industrial cities, Yanbu and Jubail. Shaw accepted the position and left for Saudi Arabia. Jordan kept in touch with Shaw, and on one of Shaw's visits back to the United States, Jordan asked him if he would be interested in working on Saudi Arabia at CSIS.

Over the next few months, Jordan and Shaw met several more times, refining further the idea of a long-term study of Saudi Arabia. During these discussions, questions arose as to how CSIS would fund such a study. Jordan suggested to Shaw that he approach Booz Allen & Hamilton because, in Jordan's words, "it had a strong interest in Saudi Arabia as a result of the contracts they held." The consulting firm, in turn, agreed to contribute $25,000 in seed money toward the $125,000 budget and the project on Saudi Arabian modernization formally commenced in late 1978, with Shaw as its director.

Another $25,000 came from the Avco Corporation, a diversified multinational company that maintains multimillion-dollar service contracts with the Saudi government to operate twenty-four Saudi civilian and military airports. Smaller grants were awarded by companies such as the Japanese-based Mitsubishi ($5,000) and Ensearch ($2,000), both of which also have business ties to the Middle East. The rest of the funding came from CSIS's pool of grant money provided by other corporations connected to the Middle East and from internally generated CSIS funds. The project set up seminars for representatives of corporations, including those that had given money, academic experts, and government representatives from Capitol Hill, the Department of Defense, and the Department of State. One of the principal participants in these seminars was David Long from the Department of State.

In the words of one of his superiors at the Bureau of Intelligence and Research at the State Department, David Long is "one of the most knowledgeable experts on Saudi Arabia in the United States." Born in 1937 in Georgia, he served as a foreign service officer at various Middle East posts including the Sudan and Saudi Arabia from 1963 through 1968. He returned to Washington where he concentrated on Saudi Arabia at the State Department. In the mid-1970s, Long took a leave of absence to serve as executive director of the pro-Arab Center for Contemporary Arab Studies at Georgetown University. During that same year Long also served as a senior fellow of CSIS, where he wrote a paper on Saudi Arabia. Known for his particularly close relationship with key Saudi officials both here and in Saudi Arabia, Long was the master's thesis adviser to the Saudi Defense attaché in Washington, Prince Bandar bin Sultan at the Johns Hopkins School of Advanced International Relations.

In mid-1980, according to State Department sources, U.S. Ambassador to Saudi Arabia John West sent the State Department a special request. He wanted a study prepared on the subject of stability in Saudi Arabia. He specifically asked that David Long undertake the study. Normally, the staffs of the embassies prepare national reports and analyses, especially if the work requires a presence in the host country. In this instance, however, Long, stationed in Washington, was assigned the task.

At the time West requested the study, the Saudis were in the process of renewing their demands for the offensive F-15 equipment. Though Carter administration officials had promised the Saudis they would press Congress to approve such a sale, the political timing was clearly not propitious. A severe case of the jitters prevailed over the entire Middle East—the "crescent of crisis" as *Newsweek* labeled the area in a splashy cover story. In late 1979, American embassy officials in Iran had been captured and taken hostage; the American embassy in Islamabad, Pakistan, had been destroyed and an American soldier killed; the U.S. consulate in Tripoli was burned by a rampaging crowd; and a religious zealots' insurrection in the holiest Islamic city, Mecca, had persisted for five weeks before the Saudi national guard, abetted by anti-terrorist French troops, subdued the rebellion. Ambassador West, a staunch supporter of Saudi Arabia, was seeking a clean bill of health for the Saudi regime. Speculated a government official who served under West, there was only one man who could do the job.

So, in September 1980, Long flew to Saudi Arabia. The trip was paid for by the State Department's Bureau of Near East and South Asian Affairs. And the staffs of the embassy, the U.S. Liaison Office, and the American Consulate General assisted Long in preparing and writing his report. By December 1980, Long had finished his work, returned to Washington, and assembled his study. The ninety-one-page double-spaced report was classified secret and distributed to various government officials under the authorization of Long's superior, Deputy Assistant Secretary Philip Stoddard.

The classified study provided an exhaustive analysis of how economic modernization has affected the Saudi society. It included, but was not limited to, an assessment of the politics within the royal family and the family's relationship to the Saudi people, the status of the military establishment, the political orientation of the students, the role of the religious leadership, the extent of dissident activity, the role of women, the magnitude and effects of corruption, and the treatment of the Shia minority. The report probed the inner workings of Saudi society with such intimate and precise details that only a person with unique access to Saudi leaders and informants could have written it. In shocking descriptions, the report detailed evidence of serious and growing problems certain to destabilize the regime—ranging from the people's mounting disaffection for the decadent Saudi princes and the Shias' bitter hatred of the Saudi leaders to the hopeless inadequacies of the military.

Besides being distributed to Long's bureau, Intelligence and Research, the report was sent to other State Department bureaus, some overseas posts, the CIA, Commerce Department,

Defense Intelligence Agency, and the Pentagon. The study was received at the State Department as an important report. Almost every top government officer who had anything to do with top-secret and high-level Middle East matters knew about it.

In January 1981 the new Reagan administration decided to accede to Saudi demands for the F-15 equipment and the AWACs. But from the moment the proposed sale was announced, it encountered stiff opposition. Beating a hasty retreat, administration strategists decided to postpone formally notifying Congress of the sale until after the summer, when it was hoped the administration could regroup and more effectively counter the opposition.

One of the main arguments against the sale revolved around the possibility that American technology would be compromised, as had occurred a year and a half before in Iran when the Phoenix air-to-air missiles and F-14 aircraft were not retrieved before the shah fell, and are believed to have fallen into the hands of the Soviets through their agents in Iran. On April 2, 1981, barely three weeks after the Reagan administration announced the sale, twelve U.S. Air Force F-15 fighter pilots sent a strongly worded and unprecedented public letter to Congressman Tom Lantos deploring the sale of the AIM-9L Sidewinder air-to-air missiles, which previously had been considered one of the less controversial components of the sale. The opposition to the transfer of AWACs technology to a potentially unstable country like Saudi Arabia was even more intense.

In promoting the sale, therefore, the administration needed to allay fears about internal ferment in Saudi Arabia, even if it meant deceiving the Congress and the public. James Buckley, as the under secretary, had taken the lead in briefing senators, congressmen, and the press—and was made aware during private and informal consultations with skeptical senators on Capitol Hill of the formidable task ahead. He therefore asked the assistant secretary of state for Near East Affairs (NEA), Nicholas A. Veliotes, to provide him with talking points and a concise briefing paper that he could use to nullify the claims that Saudi

Arabia was unstable. Veliotes, a veteran of the foreign service and a former ambassador to Jordan, assigned the task to the Arabian Peninsula Affairs desk. There, Roger Merrick, an Arabic-speaking foreign service officer, took the lead in preparing the memo for Buckley.

Merrick apparently perused Long's secret report on Saudi instability, because he used a couple of phrases from it. The end product was a concisely written four-page memo and two pages of talking points—bulleted items designed for quick reading. The memo was cleared by Merrick's superior, Deputy Assistant Secretary Joseph Twinam, a longtime specialist in the Arab world, who two years earlier had helped convince senators not to publish a report on Saudi Arabian oil production. On July 9, 1981, Veliotes sent the materials to Buckley, who soon forwarded a copy to Richard Allen, the President's National Security adviser, designated to take overall charge of the White House AWACs campaign. The papers were to be used in briefing senators and the press. Copies of the papers also filtered down to sympathetic senators for use against the anti-AWACs arguments made by their colleagues.

The briefing paper for Buckley portrayed Saudi Arabia as a phenomenally stable regime, free from any "types of dissension or organized opposition which have normally preceded revolutionary situations or serious internal instability in other Middle East countries." The Saudi regime "enjoys a strong tradition of legitimacy," the memo stated. "Cohesion is the pillar of strength within the Royal Family," the memo added, and "there are no signs of factionalism along ideological or policy lines." The memo also mentioned strong loyalty in the military. And to buttress this assertion, the report pointed to the fact that "members of the ruling family occupy key positions in the armed forces." The "200,000" people of the Shia minority are "geographically isolated," and the group "does not pose a serious political threat to the regime."

Unless political, social, and economic conditions in Saudi Arabia had undergone a miraculous transformation in the previous five months, NEA's briefing paper for Buckley was an abject attempt to spread misinformation. The assertions made in the memo were total distortions of the information in the original report; Long's study resolutely contradicted the briefing memo's proclamation of royal family legitimacy and stability. For example, the classified report stated that "the Royal Family is currently faced with the greatest challenge to its rule since the ineffectual reign of King Saud (1953–1964). The nature of the present challenge, however, is far more complicated than it was in King Saud's day . . . and virtually impervious to easy solutions." Moreover, bitter factional, personal, and policy disputes had erupted within the royal family, resulting in much friction, a "malaise" in decision-making, and explosive rivalry. "Degenerate personal lives," rampant corruption, and "an arrogance of power devoid of a sense of responsibility" characterized many of the younger princes, and this, the original report said, "has grated mightily on many Saudis."

Long's classified study warned ominously that because of government brutality and "repression," an uprising of the Shia community, "275,000" strong—an increase of 32.5 percent over the figure mentioned in the NEA memo—could easily occur within "two to five years." Moreover, the concentration of all the Shias in the strategically important oil-producing provinces—sanguinely referred to as "geographically isolated" in the NEA memo—could easily result in "devastation" to the Aramco oil installations, which produce 90 percent of Saudi oil. The Shias, who harbor an "abiding hatred for the Royal Family," constitute the backbone of Aramco's work force. Finally, Long's secret report revealed that the Saudi leaders "fear for the security of the regime over the loyalty of the military." In fact, the reason why "members of the Ruling Family occupy key positions in the armed forces"—a fact used as evidence of military loyalty in the Buckley memo—was that Saudi leaders did *not* trust the military establishment and were forced to fill positions with their own people.

The NEA memo served its purpose well. In fact, when he

gave testimony before the Senate Foreign Relations Committee on October 1, 1981, Secretary of Defense Caspar Weinberger went unchallenged when he said, "I do not share the feeling that there is any instability to the existing government." Nevertheless, State Department Arabists soon became involved in another disinformation campaign, again involving David Long's classified study. This time, however, Long himself played a critical role in a scheme sanctioned by top officials. In the first half of 1981, he submitted to his superiors at the State Department, for external publication, a long paper on Saudi Arabian modernization.

Such publication is not unusual for foreign service officers; indeed, many publish regularly in outside scholarly journals and are encouraged to do so. Long's submission, though, was unusual: with the exception of a series of meticulous changes, it was the same secret study he had written in December 1980. The changes reflected a fastidious effort to excise evidence of instability and to replace much negative material with positive assessments.

The proper and legal procedure for securing the public release of the secret report—like any other secret report—would have been to submit it to Long's superiors at the State Department for formal declassification. In this process, classified documents are carefully reviewed for release to the public after screening by the offices that have jurisdiction over the subject matter. If portions are still deemed too sensitive, they are deleted and left blank.

But this route and procedure were not followed with Long's paper. Long and his superiors at the State Department assumed for themselves a process of informal declassification: Long deleted what he deemed sensitive material, which turned out to be anything that revealed the lack of cohesion or stability in Saudi Arabia. He replaced it with new material, which diametrically opposed many of the judgments of his original classified report, and submitted the revised report as a "new" one.

Under State Department regulations governing external publishing, clearance is still required, but the process is not as stringent as for declassifying official material. Long's purportedly "new" report was distributed for review to appropriate officials, including his superiors at the Bureau of Intelligence and Research, Deputy Assistant Secretary Joseph Twinam and others at the Bureau of Near East and South Asian Affairs, and officials at the State Department's Bureau of Public Affairs.

Not one question was officially raised about the propriety or legality of Long's submission in any of these offices, although virtually all of the officials who reviewed it knew of its extraordinary similarity to, and selective replication of, the classified report.[2]

After approval was granted, Long sent his paper to his friend Shaw at Georgetown's CSIS, where Shaw had been administering the Saudi project. The ruse was completed when Shaw, after adding twenty pages of his own on Saudi development, had a draft typed. The completed draft stated that the report was written by both Shaw and Long and identified it as a CSIS paper.

At CSIS, Shaw had become involved in pro-AWACs lobbying efforts, according to CSIS sources, helping to pinpoint key senators. In mid-October 1981, he personally delivered two dozen copies of his report to the Senate Armed Services Committee and the Senate Foreign Relations Committee with the explicit intention, according to a recipient of the report, of demonstrating the stability of Saudi Arabia and thus helping to muster support for the AWACs sale. These advance copies appear to have been rushed to Capitol Hill, since they were distributed in photocopied form.

In February 1982, CSIS formally released the work as part of its "Washington Papers" series on international affairs. An accompanying CSIS press release disingenuously described it as a "major new work" written by Shaw and Long, and explained that the authors "provide the first comprehensive assessment of the economic and political implications of Saudi Arabian modernization.'' The book's preface described the study as an ''outgrowth of a CSIS project begun in 1979 that sought, sensibly and sensitively, to assess future trends in Saudi Arabia.'' Thus far, over two thousand copies have been sold or distributed—making it one of CSIS's most popular books.

The $125,000 CSIS allocated for the report—which ultimately paid for a government report that had previously been written—covered the salaries of Shaw and a research assistant as well as administrative costs. According to an official of the Avco Corporation, a prime sponsor of the project, the company was able to use it for ''enhanced marketing efforts in Saudi Arabia.'' Said the spokesman, ''it's nothing to spend a bit of money to get these enormous contracts,'' adding that ''Avco employs 11,000 people in Saudi Arabia.''

CSIS officials stated that they were pleased with Shaw's work; they were unaware that 90 percent of the report was taken from a classified government study. In an interview with me, CSIS administrators claimed that Shaw wrote ''60 percent'' of the book and that he ''rewrote the Long material'' for integration into the final manuscript.

When I inquired of Shaw how much he wrote, he responded, ''most of the report.'' Was any material taken from an earlier report written by Long? I asked. ''Phrases may have been used, that's all,'' said Shaw. He said he did not think that any of Long's material came from a classified government report.

David Long proved far more elusive. Almost two years after I first left telephone messages, he finally returned my calls in April 1984. Though he acknowledged that he had used his classified paper as the basis for the CSIS report, he said that the changes reflected ''discretionary, diplomatic deletions.'' He insisted that he had not altered the meaning or judgment of the classified paper. ''There is nothing in the first paper that was left out that could alter the first judgment. . . . I did not change one single judgment.'' How did Long get involved with the CSIS paper in the first place? Long said that ''in participating in the

[CSIS] project, they asked, 'if there was anything in your report we can use?' So I sanitized it [the classified study].'' He added that he ''got clearance all the way.''

Most State Department officials would not respond to requests for information about how the department allowed a classified report to be released without any declassification, deceptively edited, and distributed widely. The few State Department officials who did respond attempted to quash my inquiry by giving me information that was factually inaccurate. One senior official at the Bureau of Intelligence and Research, Deputy Assistant Secretary Herman Cohen, denied that any improper actions had occurred. He claimed that Long's original report was ''written as a private research paper,'' that the ''research was not financed by the U.S. government,'' and that the report was classified secret only as a preliminary step before publication in the CSIS forum. Another official, in a different explanation, said that the ''two reports were totally different reports, and therefore no improper declassification has occurred.''

The 110-page CSIS version is organized a bit differently from the 91-page State Department report. For example, the introduction to the classified study became the conclusion to the CSIS version, and some paragraphs were switched. But over 90 percent of the text of the classified study was incorporated verbatim. Except for Shaw's additions on economic development, the remaining 10 percent that was removed, truncated, or changed contains the critical revelations about internal ferment and dissent within Saudi society.

Though it is beyond the scope of this book to provide an exhaustive comparison of the two reports, the following sample excerpts demonstrate how, under the State Department's political alchemy, the alarming political observations set forth in the classified study were deceptively transformed into calm political assessments.[3] This comparison also provides a unique opportunity to observe firsthand the flagrancy of the State Department fraud and to see how the classified report was doctored.

The classified State Department study reveals that questions about the political legitimacy of the royal family, which is determined by its acceptance by the citizens of Saudi Arabia, has reached crisis levels. There is a widespread perception in Saudi Arabia of "personal degeneracy and moral corruption" in the royal family. The CSIS version, however, belittles the issue of corruption and actually substitutes false material.

ROYAL FAMILY INSTABILITY

STATE DEPT. STUDY*

The greatest destabilizing aspect of corruption in Saudi Arabia is probably psychological. The practices themselves are not unequivocally condemned, indeed they comprise the national pastime.

On the other hand, the perceived extent of Royal Family involvement in corrupt practices has served to undermine the reputation for strict Wahhabi integrity which has over the years been a hallmark of the regime. Conspicious consumption and high spending life styles in the watering spots of Europe and America have contributed to the image of personal degeneracy and moral corruption among members of the Royal Family, particularly the younger princes. This image, accompanied by one of arrogance of power devoid of a sense of responsibility, is displayed by many of the younger princes in business and other activities, and has grated

CSIS VERSION

The greatest destabilizing aspect of corruption in Saudi Arabia is probably psychological. The practices themselves are not unequivocally condemned; indeed they comprise the national pastime.

In general, members of the royal family are, if anything, more circumspect in their business dealings than most of the business community; appearances in this respect are central to their sense of legitimacy. If the royal family were to be perceived as "excessively" involved in corrupt practices, this could serve to undermine the reputation for strict Wahhabi integrity that has over the years been a hallmark of the regime.

*Material removed or changed in the CSIS version is printed in italic type.

mightily on many Saudis. It would be far easier for politically disaffected Saudis to accuse the regime sullied with the reputation of corruption of being the cause of their frustrations than to look to the real though complicated root cause—the impact of technology and modernization on a traditional Islamic society (pp. 83–84). [These page numbers refer to the pages of the original documents.]

It would be far easier for politically disaffected Saudis to accuse a regime sullied with corruption of being the cause of their frustrations than to look to the real, though complicated root cause: the impact of technology and modernization on a traditional Islamic society (p. 54).

[The CSIS report then included additional material to discredit reports of corruption. The new material pins the blame for corruption becoming an issue on "opponents" of the regime.]

On balance, the issue of corruption in the royal family stems from a few well-publicized incidents involving a small number of princes who have already been reprimanded by the family. The publicity from these few examples has, however, been juxtaposed with the immense sums inherent in the development process and recycled by opponents of the regime to suggest continuing malaise or instability in the kingdom (p. 54).

ROYAL FAMILY CORRUPTION

STATE DEPT. STUDY

Not all Saudi princes (an estimated 3,000 males) are interested in public service, and many, such as the sons of Crown Prince Fahd, have *unsavory reputations* for sharp business dealings and *for degenerate*

CSIS VERSION

Not all Saudi princes (an estimated 3,000) are interested in public service, and many have gone into business. Some princes have gained reputations for sharp business dealings, and the public *personalities. The public image of the Royal Family is thus of major concern to the leadership and may ultimately be its undoing. In the absence of strong leadership and with access to so much money, it will be difficult to prevent the family image from being tarnished. In addition, along with the influx of such large sums of money, some members have developed an arrogance of power without a commensurate sense of social responsibility, and this has undermined the family reputation. It does not as yet appear to be a major source of political instability, and it may not become so as long as everyone else has the opportunity to make a fortune, albeit on a lesser scale. Nonetheless, the family image could in time cause widespread disaffection should the real social and political causes of frustration appreciably increase.* How the family deals with its errant members will, therefore, be increasingly important, particularly so long as the present collegial leadership is in power (p. 12).

image of the royal family has become a major concern to the leadership (p. 63).

How the family deals with its many members and with their varied interests will therefore be increasingly important, particularly as long as the present collegial leadership is in power (pp. 63–64).

Another threat to the stability of the royal family, states the classified study, is the smoldering resentment of the Shia minority, who have been kept as second-class citizens. Yet their relatively small numbers out of Saudi's seven million people mask the disproportionate influence they bring to bear on both the cohesion of Saudi Arabia and the safety of continued oil production. Located primarily in the eastern oil-producing province, the Shias constitute at least one-third of the Aramco work force and are responsible for pumping 70 percent of all Saudi oil. The Shia brand of Islam is more fundamentalist than the predominant Sunni sect. (The Shias are the dominant group in Iran and were led by the Ayatollah Khomeini in his toppling of the shah in 1979.)

Removed from the CSIS version is the report of the frightening degree of Shia antipathy toward the Saudi authorities, including the possibility that they "could pose a major security threat over the next two to five years [which] could be devastating to oil production."

THE SHIA MINORITY

STATE DEPT. STUDY

CSIS VERSION

Shia Antipathy:

Although the visit of the King to the Eastern Province in late November 1980 had an immediate and positive effect, *one cannot totally reverse Shia antipathy toward the government amounting in some cases to deep and abiding hatred for the Royal Family which years of neglect had bred* (p. 55).

This [Shia antipathy] was reversed during the visit of the king to the Eastern Province in late November 1980, which had an immediate and positive effect, but one cannot totally negate years of neglect (p. 99).

The King's Visit:

. . . The best received act of the government, however, immediately preceded the visit when several hundred Shia *political* prisoners were released in time for Ashura [a Shia holiday]. *Apparently, the government made a deal with Shia leaders to release prisoners in return for there being no disturbances on the Shia religious holiday* (p. 55).

best received act of the gov —

The ~~key question for the future~~, ernment, however, immediately preceded the visit when several hundred Shias were released from jail in time for Ashura (p. 99).

Anti-American Agitation:

Partly as a result of Iranian agitation, there is a good likelihood that at least some Shias will, for the first time, begin to associate their grievances with the U.S. and, by extension, with Aramco as well as with the Saudi regime. Anti-regime and anti-American literature has recently begun to appear at Aramco and elsewhere in the Eastern Province, and there are reports of Shia dissident organizations being established (p. 58).

Security Threat to Saudi Regime:

In the longer run, *continued insensitivity by the regime to Shia grievances and the steady build-up of disaffection among the younger generation could pose a major security threat over the next two to five years. Of course, reaction to police brutality or to destabilizing external events could bring on a threat much sooner. Moreover, the consequences of possible Shia uprisings, no matter how remote at present, could be devastating to oil production. Skilled Shia workers are located at some of Aramco's most sensitive installations— those where many non-Saudis are prohibited for security reasons— and thus ideally situated to conduct acts of sabotage. The older generation is probably too loyal to Aramco to consider such acts although loyalty might decrease as Aramco becomes Saudi.*

In the longer run, the effectiveness of recent government policy to increase substantially public services in the Shia areas could be crucial in defusing Shia frustrations, particularly among the younger generation (p. 100).

The critical question for the future, therefore, is which way the younger generation will go (pp. 58–59).

The key question for the future, therefore, is which way the younger generation will go (p. 100).

Another area that was radically altered in the CSIS report involves the royal family's fear of the military and the Saudi students. The State Department report makes some startling observations: that the royal family does not fully trust its own military establishment; that expatriates are increasingly being relied on to fill major "support and logistics roles" because of a lack of a "work ethic"; that the military cannot fully absorb all its equipment; and that Saudi geology students are not even trusted with information about the oil fields. All of this was deleted in the CSIS version. Moreover, the State Department report also portrays the Saudi military establishment as unwanted by the royal family but required by national "self-respect"; the CSIS version attaches a different motivation—one that enhanced the Reagan administration's pro-AWACs campaign.

FEARS AND PROBLEMS WITHIN THE SAUDI MILITARY AND AMONG STUDENTS

STATE DEPT. STUDY

CSIS VERSION

Royal Family Fears of Military:

The Saudis have long been ambivalent about the efficacy of a modern military establishment. On the one hand, the Royal Family is acutely aware that military establishments in the Middle East tend to overthrow monarchies, and there is *real fear for the security of the regime*

The Saudis have long been ambivalent about the efficacy of a modern military establishment. On the one hand, the Royal Family is acutely aware that military establishments in the Middle East have often overthrown monarchies, and there over the loyalty of the military. *The arrest of military officers, particularly from the air force, in 1969–1970 for subversive activities greatly reinforced Saudi distrust in a professional military establishment. Traditionally, the Saudi military has never been issued fuel and ammunition in large quantities or at the same time. Moreover, younger members of the Royal Family are encouraged to enter military services, particularly the air force* (p. 15).

Saudi View on Military Establishment:

On the other hand, the Saudis have become convinced *that no self-respecting Middle Eastern state with the economic power Saudi Arabia possesses can avoid the necessity of organizing and training an effective modern military establishment* (pp. 15–16).

Military Loyalty:

In addition to the Royal Family reluctance to trust too greatly in the loyalty of the military establishment, and hence the proclivity to slow down and delay progress in many programs, the Saudi military faces the same extreme manpower constraints which are endemic also in the civilian sector (p. 16).

Quality of Military:

Military training and arms purchases are subject to several additional constraints. Due largely to

is natural concern over the loyalty of the military (p. 66).

On the other hand, the Saudis have become convinced that to protect their vast oil and economic resources, it is important that they organize and train an effective, modern military establishment (pp. 66–67).

The Saudi military faces the same extreme manpower constraints that are endemic in the civilian sector (p. 67).

Saudi ambivalence toward a modern military force, there is relatively little to show for nearly 30 years of U.S. military training and British training before that (p. 17).

Currently, many of these positions are being filled by contract expatriates. In the case of hostilities, however, expatriates cannot be depended upon and the Saudis lack the experience. *The problem is exacerbated by the general lack of a work ethic in Saudi Arabia which results in Saudi officers and men being perfectly willing to allow expatriates to fill major support and logistics roles, even when the Saudis have the ability to do it themselves (p. 18).*

The Navy is by far the most neglected branch of service. Most of its senior officers are from the Hijaz and Hijazi officers are generally distrusted by the Najdi [central Arabia] oriented regime. Finally, the French arms purchase is expected to place an almost impossible burden on the already strained manpower resources of the navy (pp. 19–20).

Distrust of Students:

The University of Petroleum and Minerals has prestige in science and engineering, but has a reputation for *radical student* views. *Ironically, its petroleum engineering students study Oklahoma oil fields because they are not trusted by the government with information about Saudi oil fields. (p. 33).*

Currently, many of these positions are being filled by contract ex patriates. In case of hostilities, however, expatriates cannot be depended upon, and Saudis lack the experience (p. 68).

The University of Petroleum and Minerals is noted for science and engineering but has a reputation for outspoken students (p. 83).

Had the State Department report's shattering portrait of the Saudi kingdom been made public at the time of the AWACs debate, the evidence of instability and corruption might very well have doomed the sale of AWACs to Saudi Arabia.

What was especially remarkable about the ruse was not just the brazen duplicitous treatment of Congress, but the multiple levels of vested interests that ensured the success of the deception.[4]

For years, the fear of offending Arabia and of making the American public skeptical of Saudi Arabia's reliability or acceptability as an American ally has led to a far more subtle approach by State Department officials attempting to shape public opinion. They are unwilling to make public any criticism of Saudi Arabia, even at the cost of blatantly compromising human rights standards or, worse, releasing disingenuous and half-true material to the press and public.

One need only glance at the section on Saudi Arabia in the annual State Department human rights survey to see a meticulous effort to portray human rights abuses sympathetically as the product of indigenous cultural and religious mores. Somehow, even after reading references to beheadings, stonings, "severances of the hand," lack of habeas corpus, inequality for women, and a "strong emphasis on obtaining confessions," one still gets the impression that Saudi Arabia is a bastion of democracy. This assessment is made even more offensive by a comparison to the appropriately critical State Department evaluations of human rights violations in Poland, the West Bank, and Zimbabwe.

The often brutal treatment of American citizens incarcerated in Saudi Arabia is a major scandal, but neither the American embassy in Riyadh nor the State Department has ever issued any public criticism of Saudi arrests—of which there have been many—or even commented on this as a general problem. (An even greater problem encountered by Americans arrested in Saudi Arabia is the general indifference to their plight by the American embassy and consulates.)

Other attempts to color the public debate have also revolved around the selective use of information. While I served on the Senate Foreign Relations Committee, State Department Middle East analysts would often be invited to brief the staff and senators. In February 1980 I helped organize a set of closed-door hearings on the "political, social and economic factors affecting the Persian Gulf countries." Over a three-day period, State Department officials, in addition to CIA analysts, provided material and testimony on factors affecting Saudi oil production, focusing on internal Saudi instability and, to a lesser extent, on the Arab-Israeli dispute.

When the State Department officials were asked by the committee staff to declassify their testimony so that the hearings could be made public, the officials declassified all references to the Arab-Israeli conflict and Israeli policies as factors affecting Saudi production; but material relating to internal Saudi instability, such as Islamic dissidence, remained classified. A reading of the selectively declassified hearings would have inaccurately suggested that the Arab-Israeli conflict was the cause of reduced Saudi oil production—a conclusion that State Department analysts did not support behind closed doors. State Department officials were then told that unless they were willing to declassify everything, there would be no partial declassification. In the end, these hearings were not published, because State Department officials refused to declassify all of their testimony.[5]

Through the selective use of information, State Department Arabists have been able to ensure that the underpinnings of the U.S.–Saudi relationship remained secure and protected from any critical congressional or public scrutiny. But an even greater problem emerged around the role played by State Department officials after they retired from office. The opportunity to exploit their previous positions and become middlemen for petrodollar investment proved irresistible.

13

THE REVOLVING DOOR

There is one thing I do personally. I let no Zionist statement go unchallenged.

Andrew J. Killgore,
former U.S. ambassador
to Qatar

Once out of office, many former ambassadors to the Arab world reverse their traditional roles and become goodwill emissaries in the United States on behalf of Arab governments.

The American ambassador to Syria, Talcott W. Seelye, did just that. He called reporters into his Damascus embassy office for a final press conference upon his retirement to make his views clear. The tall, lanky fifty-nine-year-old ambassador was concluding thirty-two years in the foreign service—most of them in the Arab world. Seelye was born in Lebanon to American parents; his father had been president of the American University of Beirut. The comments Seelye made to reporters on August 31, 1981, when he retired, appeared to reflect the cumulative effect of his family background and the twenty-eight years he served in Arab countries or Arab country desk positions in the U.S. State Department.

Sitting in an easy chair as he puffed on a cigar, Seelye called on the United States to abandon the Camp David Accords because the agreement had turned into a "red flag" to other Arab governments. He then urged the Reagan administration to commence a "dialogue" with the PLO, without requiring the PLO to renounce terrorism or to recognize Israel.[1] The ambassador proceeded to bitterly denounce Prime Minister Begin and Israeli policies, especially the "plantation settlements" on the West Bank.

Seelye's parting blasts at the policies of the United States and Israel received extensive media coverage in the United States and the Arab world. The *Washington Post,* for example, relayed the substance of Seelye's remarks under the headline "Retiring U.S. Envoy Urges Links to the PLO." The following day, however, a State Department official disavowed Seelye's comments, saying they did not represent official U.S. policy.

Two months later, in late October 1981, Seelye's name was again cited in an international wire service story. According to the Associated Press, Seelye had been quoted in a Beirut weekly, *Monday Morning,* which was widely distributed throughout the Middle East.[2] The former ambassador, in a telephone interview, had praised the eight-point Saudi peace plan. The plan—which did not call for explicit Arab recognition of Israel, but did demand that Israel immediately withdraw from all of the occupied territories—could "very definitely serve as the basis for a new peace plan formula."

By the following year, Seelye had emerged as a staunch advocate of the Arab position, popping up at various pro-Arab lobbying organizations, business groups, and public policy institutions. Appearing, for example, before the May 1982 annual convention of the National Association of Arab Americans, which included top Arab government officials among its guests, Seelye urged the United States to negotiate with the PLO, adamantly asserting that the PLO no longer condoned terrorism. To the contrary, he insisted, the PLO "rejects" and "forswears terrorism." As evidence, Seelye cited several examples, such as the "protection" afforded Seelye by the PLO on his mission to Lebanon as President Carter's special envoy following the 1976 assassination of Ambassador Francis E. Meloy, Jr., and the "important effort" waged by the PLO to gain the release of the American hostages in Iran.[3]

At another point Seelye argued against exclusive Israeli control of Jerusalem. But his rationale revealed a peculiar view of the Jewish claims to Jerusalem: "Among other things, we will have to convince Israel that the Holy City of Jerusalem cannot be forever controlled exclusively by the *smallest* and *least powerful* of the religions for which it is holy" (emphasis added). And he was moved to "observe that a strong American president can override a domestic lobby in the pursuit of U.S. national interests."

Seelye's animus toward Israel was clearly in evidence that night of October 14, 1982, at a debate in New York City sponsored by the Amherst University Alumni Association in New York. Held in the Church Center, part of the U.N. complex, the debate pitted Seelye, a 1947 graduate of Amherst, against Amherst history professor Gordon Levin, Jr., before an audience of over one hundred alumni and parents.

"I have been out of the U.S. State Department for a year," Seelye said before beginning his formal statement, "which is the only reason I can make these candid comments today." Though Seelye issued a number of criticisms of previous Arab policies, he reserved the bulk of his comments for Israel. He proceeded to deliver what the Amherst student newspaper characterized as a "blistering attack" on Israeli Prime Minister Menachem Begin and the policies of Israel in language and virulence that shocked several participants.[4]

Condemning "unprovoked" Israeli "aggression" in Lebanon, Seelye charged that Israel had killed "15,000 men, women, and children" and wounded "four times that number" in an effort, he said, "to deal a death blow to the Palestinians." Another example of Israeli "aggression," said Seelye, was the Israeli attack on Iraq's nuclear reactor, which was totally unjustified, because Iraq was "no way near developing the atomic bomb capacity and was open to international inspection."

Seelye's outrage at Israel increased dramatically as the evening progressed. He contended that "we can't get any military cooperation between the U.S. and Arab states as long as we have a close military relationship with Israel." And in response to his opponent Levin's statement that the Reagan administration "realized that Israel had legitimate aims in entering Lebanon to defeat the armed forces of the PLO," Seelye characterized Israeli Defense Minister Ariel Sharon as "indistinguishable from a Nazi stormtrooper." At another point Seelye asserted that the U.S. government had exempted American Jews from paying tax on Israel Bonds—development bonds for the state of Israel. He made that claim while arguing that leverage be applied to American Jews, such as withdrawing the "tax-exempt" status of these bonds. The former ambassador also called for requiring "certain Jewish groups who lobby for Israel to register as foreign agents."[5]

On these issues, and on several others, Seelye was heatedly challenged by a distinguished alumnus in the audience, Robert Morgenthau, district attorney of New York County and a fraternity brother of Seelye. "I object to your very misleading statements," Morgenthau said as he rose to deliver a stern rejoinder. He said that southern Lebanon had long been used as a terrorist staging ground for the PLO; that contrary to Seeyle's assertion, American Jews are required to pay taxes on Israel Bonds; that American Jews are not extensions of the Israeli government; and that it would not be appropriate for the American government to pressure American Jews for the purpose of changing another government's policies any more than it would be proper to pressure other ethnic groups.

The views Seelye expressed before the Amherst alumni and parents were not unusual for him. Articulate, self-assured, and, most important, the beneficiary of a lifetime of invaluable policy experience, Seelye was well on his way toward becoming a widely sought-after commentator in the media and various foreign policy organizations. In the wake of the Israeli invasion of

Lebanon in June 1982, and the media attention devoted to it, Seelye's popularity skyrocketed. He was featured on numerous television shows, such as The MacNeil-Lehrer Report, "Nightline," "Good Morning America," and "Today." And he wrote several opinion pieces in which he expanded on his criticism of Israel and his virtually blind embrace of Arab policies.

In one opinion piece, published in the *Washington Post* and later reprinted in the *International Herald Tribune* and the *Middle East Times* published in Cyprus, Seelye passionately argued that "no country in the Middle East is more misunderstood than Syria."[6] There was no truth, Seelye claimed, to the popular image of Syria as "violently anti-American" or as a "Soviet satellite," "Qaddafi-like in its rejection of Middle East peace" and "eager to go to war with Israel." Rather, Syria's bad image stemmed from "a combination of Syrian declaratory extremism and anti-Syria propaganda." Assad, in fact, was "fundamentally a political moderate posing as a radical," according to Seelye.

Missing from Seelye's analysis were some salient and indisputable facts that would have provided specific reasons for Syria's poor image: the documented role Syria played in abetting international terrorism (resulting in that country's being designated by the U.S. State Department as one of only five terrorist-supporting countries in the world); Syria's sponsorship of terrorist attacks on American forces in Lebanon; and most important, President Assad's ruthless human rights record. (Besides torturing hundreds of political prisoners, documented by Amnesty International, he was responsible for the massacre of as many as 20,000 Moslem citizens in Hama, Syria, in 1982.)

At the end of the *Washington Post* article, was a simple one-sentence description of Seelye's background: "The writer was ambassador to Syria from 1978 to 1981." And in an article Seelye wrote for the *Christian Science Monitor*—in which he justified Syrian attacks following the 1948 War on Israeli farmers in the Golan Heights—he was again identified only by his foreign service background.[7] There was nothing inaccurate in either of those bylines; they simply did not reveal what Seelye was doing at the time he wrote the article.

Nor were the nationwide audiences who saw Seelye on television news shows informed of the exact nature of Seelye's occupation. If anything, the ambiguous descriptions may have enhanced the public's perception of Seelye's objectivity. On "The MacNeil-Lehrer Report," for example, Seelye was interviewed four times between June 1982 and January 1984—generally defending Syrian actions, praising Assad, and criticizing Israel. The introduction that preceded each of his appearances included his experience in government, particularly his position as a former ambassador. But his current occupation was identified as "consultant on Middle East affairs"; "international consultant here in Washington, specializing in Syrian or other Middle East matters"; and a "Washington-based consultant on international affairs."[8] If the exact nature of Seelye's current professional endeavors had been revealed, audiences might have concluded that Seelye was not exactly a disinterested observer.

Because, by early 1982, months after leaving the foreign service, Seelye had signed up as a consultant with Rezayat America, Inc., a Saudi-owned company based in the United States. The firm was part of the Alireza Group, a multibillion-dollar Arab conglomerate of numerous trading companies that have teamed up with American and European multinationals to handle multibillion-dollar industrialization service contracts, such as pipeline construction and cargo handling, in Saudi Arabia, Kuwait, the United Arab Emirates, and Oman. Rezayat America, which represents American companies exporting to the Persian Gulf, retained Seelye for three days a week and provided him with an office in the firm's Fifth Avenue, New York City headquarters. Seelye had been asked to help set up Rezayat's office in New York by Abdullah Alireza, one of the firm's owners and a long time friend of Seelye.

Ten months later, Seelye left Rezayat America and became the director of Middle East Services for the Advest Group, a growing investment and financial securities firm, headquartered in Hartford, Connecticut, with offices throughout the country. A special company brochure sent to prospective and current clients touts Seelye's special attributes, such as the fact that "he maintains close ties with the most senior government officials in the Arab countries of the Middle East." For $20,000 a year in commissions, Advest clients can take advantage of a "first-team all-American expert," who—Advest boasts—is concerned about the "inaccurate perceptions being passed on to investment decision-makers." Clients receive "confidential" quarterly reports from Seelye, a "Mid East Alert" telex should "a momentous development" occur, "daily access" to Seelye by telephone, and private briefings by Seelye in the client's office.

Seelye's access to the Arab world and its leaders is especially critical in an additional line of work for Advest: Seelye leads Advest's top American investment financial counselors on "fact-finding tours" of various—often oil-rich—Middle East countries. On these tours, which cost clients either $12,000 up front or $20,000 in subsequent brokerage commissions, Seelye helps arrange meetings with "high ranking government officials, U.S. embassy representatives, and local senior business executives."[9]

Prior to one trip, for example, scheduled for October 1984, Advest sent letters to clients in June 1984 inviting them to accompany Seelye on a thirteen-day tour to Syria, Iraq, North Yemen, and the Sudan. On this trip, Seelye's good relations with the Syrian president were considered crucial. Advest wrote that "based upon his life-long contacts, he is able to schedule appointments with high-ranking government officials such as President Assad of Syria and oil, industry, business, and banking leaders."

Advest itself has had considerable experience in sponsoring these "fact finding" trips, having conducted seventeen such business trips to the Middle East prior to Seelye's arrival. One trip, however, in 1975, proved to be a bit sticky when it was disclosed that Advest advised its travelers to bring a "signed statement by a clergyman attesting that the participant is a Christian.''

Another former ambassador who has parlayed his connections and views into a lucrative consulting practice is James E. Akins, whose vehement support of the Arab cause occasionally makes him appear more pro-Arab than the Arab officials. One episode revealing Akins's zeal occurred during the last week of September 1981 in London. The occasion was the annual energy conference, co-sponsored by the *International Herald Tribune* (owned by the *New York Times*, the *Washington Post*, and Whitney Communications Corporation) and the *Oil Daily*, a widely read energy newspaper. In attendance for the two-day conference, which cost each participant $825, were two hundred bankers, businessmen, diplomats, and journalists. Akins and Saudi oil minister Yamani were among the featured speakers. Held at the Royal Garden Hotel, the conference coincided with the hotly contested American debate over the American sale of AWACs to Saudi Arabia.

At one much heralded session, Yamani spoke about his country's oil policies. He explained that Saudi Arabia's desire to keep the price of oil from rising too steeply stemmed from economic self-interest—the need to be sure of long-term exports. And in order to stimulate world economic recovery and thus increase long-term demand for oil, Yamani said that his country wanted to freeze the price of oil. At the conclusion of his talk, one participant interviewed by the *International Herald Tribune*'s Joseph Fitchett noted, ''You could hear those corporate planners mentally shelving their planned investments in alternative fuel sources as Sheikh Zaki reassured them about Saudi moderation.''[10]

After Yamani finished his speech, it was Akins's turn—but suddenly Akins discarded his written statement and delivered his own impromptu lecture. Why? Because Yamani left an impression on the audience that the United States did not need to please Saudi Arabia—and thus did not need to sell the AWACs—insofar as Saudi Arabia's moderate oil policies were a function of economic self-interest.

Akins explained to the audience that Yamani's statement was "just not the whole story." And what was "the whole story"? Akins enlightened the crowd with his own interpretation of Saudi oil policies: "[I]t is hardly possible that the oil policies Saudi Arabia has followed could be in Saudi Arabia's own narrow political or economic interest." Rather, Akins insisted, "the most important reason for the Saudi position is a political one; that is the United States had asked Saudi Arabia to produce more oil, to hold down oil prices, and to defend the dollar. The Saudis' response has been consistently and dramatically positive in all fields." Even at the cost of destabilizing internal and external political strains, Akins claimed, Saudi leaders have maintained higher oil production levels—out of friendship for the United States and an expectation that the United States would reciprocate politically.

In his talk, Akins launched into a condemnation of Israel and also warned that the Saudis would be forced to "react" by cutting oil production if the sale of AWACs did not go through: "They will have to react because they will not be taken seriously if they don't, and the only place they can react so we will notice is oil policy." Akins's prediction of Saudi oil sanctions was all the more remarkable since Yamani, in his speech, had specifically disavowed any link between AWACs and his country's oil policies. The former ambassador then outlined for the audience a series of specific punitive actions the Saudis might take, specifically citing the possibilities of reimposing an oil production ceiling of 8.5 million barrels a day and reducing investment in the United States.

But Akins's cry of wolf struck one participant as a bit hyperbolic. If the Saudis, Akins was asked, were so sensitive to American policies, why had they not long ago "reacted" adversely to American positions that were clearly hostile to their interests? Akins could only respond that the Saudi leaders had been "too timid" in the past.

At the end of his talk, Akins flatly predicted that the "Saudi leadership is going to be convinced, as most Saudi citizens are, that the oil policy has not worked and must be changed, despite what Sheikh Zaki has said here.'' Sheikh Ahmed Zaki Yamani, noted the *International Herald Tribune,* ''did not appear to be upset being publicly corrected in this way.'' In his prepared comments, Akins had attacked the ''enemies'' of Saudi Arabia, specifically singling out ''Jewish writers'' Joseph Kraft and William Safire for their ''racism''—comparing their views of Arabs to the way Nazis viewed Jews.[11]

The fact that Akins felt compelled to try to convince the audience of the political link implicit in Saudi oil policies, even though Yamani had said otherwise, was not surprising to those who had followed Akins's career. He was fired by Henry Kissinger from his post as ambassador to Saudi Arabia in late 1975 because of his excessive partiality to Saudi Arabia.[12] Indeed, he exhibited such partiality after being on the job less than seven weeks after he was appointed in September 1973.

On October 25 of that year, only days after Saudi Arabia had imposed the oil embargo, Akins contacted Aramco executives in Saudi Arabia, urging them to ''use their contacts at highest levels of U.S.G. [United States government] to hammer home the point that oil restrictions are not going to be lifted unless political struggle is settled in a manner satisfactory to Arabs.''[13] His action was truly extraordinary. Here he was, the American ambassador to Saudi Arabia, attempting to reinforce the Arabs' blackmail of the United States. When I asked Akins about this episode in an interview with him, he said, ''They [the oil companies] should have done something. The oil companies know what American interests are.''

Akins entered the foreign service in 1953 at the age of twenty-seven, later serving in Syria, Lebanon, Kuwait, and Iraq before rising to his next-to-final position as the top State Department expert on energy matters. During his term in government, Akins was known for his brilliance, his fierce independence, and his occasionally abrasive style. After his sudden departure from government in late 1975, Akins signed up as a consultant to various American multinationals, many with business ties to the Middle East. He has continued in that capacity to the present, having made frequent trips to Iraq, Saudi Arabia, and other countries, opening doors for his clients and attempting to get them additional petrodollar business.

At the same time, Akins has maintained a high public profile, propagating his views in newspapers, before business groups and Arab organizations, on television, and in testimony before Congress. Sometimes the views he expresses, like his talk at the 1981 energy conference in London, are so zealously pro-Arab that he criticizes or attacks Arab governments for not being tougher with the United States. In a 1981 lecture at Saudi Arabia's University of Petroleum and Minerals in Dhahran, Akins told the 250-member crowd that "the Arab countries should get something for what they give."[14] He then listed the reasons why "there should be an outpouring of American gratitude to the Saudis."

Speaking before southern businessmen in Birmingham, Alabama, at a conference sponsored by a pro-Arab organization, in March 1983, Akins harshly attacked Israel as the fundamental cause of instability in the Middle East for the past thirty-five years. Then he recounted episode after episode of American "humiliation" of the Arabs—for example, the passage of the anti-boycott legislation. Akins bitterly lamented the lack of an Arab "response" to the United States.[15]

In March 1979, right after the historic Camp David Accords were signed at the White House, Akins, in an article for *Newsday,* warned that as a result of the Accords, "we must be prepared for another and wide Mid-Eastern war which could have disastrous consequences for the United States and the world."[16] Readers of Akins's commentary were informed only that he was "a retired U.S. Foreign Service officer," a "leading State Department Arabist and oil expert," and "U.S. Ambassador to Saudi Arabia from 1973 to late 1975."

In a *New York Times* interview, which appeared in September 1983, Akins was asked why he was so "certain" there would be a "major political disruption occurring in the Middle East?" Because, Akins responded, the "growth of anti-American feeling in the region has been staggering since the Israeli invasion of Lebanon."[17] Like so many times in the past, Akins's cry of wolf has proven false; Arab governments have not cut off trade with the United States nor has there been any new war. But his scare tactics have been helpful in a more personal way: his popularity in the Arab world has soared.

Another former ambassador highly regarded by Arab regimes is Andrew I. Killgore. On February 21, 1984, the tall white-haired speaker with the heavy southern drawl flew directly from Washington, D.C., to Bloomington, Indiana, to deliver a lecture at the University of Indiana. Killgore's talk, on "U.S. Foreign Policy in the Middle East," was sponsored by the university's department of Middle East studies. Killgore was billed as a former American ambassador to the Persian Gulf state of Qatar who had a long and distinguished record of government service.

In his talk on that Tuesday evening, Killgore delivered a rather circumscribed address on American foreign policy in the Middle East: he focused only on Israel, particularly its invasion of southern Lebanon.[18] The one-sided perspective was evident not only from Killgore's repeated criticism of Israel, but from the absence of any comments critical of Arab actions or policies.

He defended Syria's violation of the May 1983 Israel-Lebanon disengagement accord, which had been drawn up by Secretary of State George Schultz. He accused the Israelis of arming the Christian forces in Lebanon, though he said nothing about the fantastic number of weapons—enough to equip brigades—that the PLO had amassed in southern Lebanon, a substantial number of which had come from or were paid for by Saudi Arabia. He clearly intimated that the invasion of southern Lebanon was partly designed to capture Lebanon's water supplies, which he said the Zionist leaders had coveted since the 1920s. And he told the students that he had found "ironic the situation that the State of Israel [was] overwhelmingly dependent on what is essentially a Christian nation."

In his summation, Killgore offered some final thoughts, after forty-five minutes of berating Israeli policies: "Our interests are beginning to suffer now in a traumatic way, particularly the three hundred [marines] dead.'' In the same breath, he bid the audience ''not to forget'' the deaths of the president of the American University of Beirut and the head of the Sinai Support Mission [who was shot by terrorists in Italy]. The implication was clear: Israel was responsible.

What most students did not know is that after his retirement in August 1980, Killgore became the head of a corporation specializing in providing ''advisory services relating to business matters in the Middle East'' including ''general investment'' and ''public relations.'' Formed in April 1981, eight months after he left the foreign service, Killgore's Amrok Corporation was organized to act, according to its articles of incorporation signed by Killgore and on file in the District of Columbia's Office of Recorder of Deeds, ''as agent or representative for any individual, business concern, firm, partnership, association, corporation, or government, including any *foreign government*'' (emphasis added).

During the question-and-answer period that followed Killgore's address at Whittenberg Auditorium, one student, aware of Killgore's affiliation with Amrok, asked if Amrok had been paid by Arab governments for its services, which, he pointed out, included ''public relations.'' Immediately, Killgore shot back, ''No PR. The PR part is wrong.'' But he acknowledged serving as a consultant to American multinationals on the Middle East. A twenty-year-old junior, Seth Eisenberg, a member of a campus pro-Israel support group, then waved a copy of the deed, telling Killgore, ''That's what it says in your [incorporation] deed.'' Killgore responded, ''Who knows who put that thing out!'' Later, after the question period was over, Eisenberg and several other students approached Killgore showing him a copy of his (Killgore's) signature on the deed. ''It must be a forgery,'' the former ambassador said. But, in fact, the deed was undeniably authentic, his signature was genuine, and ''public relations'' was definitely one of the services Killgore offered.

This may have been one of the few times Killgore was

publicly confronted with his vested ties to the Middle East, and the disclosure was obviously embarrassing. But for three years prior to this event, Killgore, like other former ambassadors to Arab countries, had articulated decidedly one-sided views on the Middle East at various public forums and in periodicals, while at the same time operating his Middle East consulting business.

Killgore's criticism of Israel and Zionism has sometimes bordered on the extreme. He has aligned himself occasionally with fanatics obsessed with "Zionist influence" and its alleged conspiratorial manipulation. In 1982, for example, he signed his name to a manifesto that endorsed a PLO-advocated "democratic and secular state" to supplant Israel, and condemned "Zionist influence with the United States which has colored all news coverage and discussion about the Middle East for three decades." [19]

At a Washington meeting in mid-1982 of the Holy Land State Committee, an organization dedicated to "liber[ating] the United States from the domination of Zionism"—and which enjoys the political support of the right-wing Liberty Lobby—Killgore decried the influence of "Zionists" in the American media and boasted: "There is one thing I do personally. I let no Zionist statement go unchallenged." [20] And at another meeting of the same group in August 1983, Killgore, according to the *Spotlight*, the newspaper of the Liberty Lobby, told the audience, "My status as a Christian and as an American is threatened by Israeli actions." [21]

Born in Alabama in 1919, Killgore entered the foreign service in 1949 and was assigned to work with refugees from wartorn Europe. In 1955, however, the State Department sent him back to Washington to learn Arabic; and for almost all of his next twenty-five years, Killgore served in the Moslem world or as State Department desk officer in charge of Arab countries. He has had close contact with Arab leaders since 1964 when he accompanied Jordan's King Hussein on a tour of the United States.

On August 9, 1977, Killgore was sworn in as ambassador to Qatar. His tenure lasted almost three years. Within five weeks after he left the diplomatic post, he had organized the American Citizens Overseas Political Action Committee, a PAC designed apparently to fund pro-Arab candidates running for Congress. Killgore intended to open a bank account in Riyadh, where the PAC's treasurer resided, but the Saudi authorities denied permission for the account.[22] Still, by November 24, 1980, Killgore had received $1,750 in contributions. With the exception of $115 that he and his wife gave, all of the money came from Americans working in Saudi Arabia, including a donation from American Ambassador John West.

The following April, Killgore formed the Amrok Corporation, installing himself as head and his family members as the board of directors. Later in that year, Killgore became the first president of the newly formed American Educational Trust, a well-heeled pro-Arab "educational" organization headquartered in Washington. The organization is composed primarily of retired State Department officials and academicians specializing in the Arab world. Its activities have been directed toward promoting the views of the Arab countries, particularly on the Arab-Israeli dispute, in Washington and across the country. All six of its "white papers" on the Middle East focus on Israel, offering "exposés" and "analyses" of "Israel's pursuit of Arab water resources," "Zionist mythology," and Israeli "aggression."

In another new, equally well-endowed, and politically predisposed group with which Killgore is affiliated—the American Arab Affairs Council—he was joined by eight other former American ambassadors to Arab countries. Of these eight, at least five including Talcott Seelye have become involved in the recycling of Arab oil money in their post-diplomatic service. William Stoltzfus, Jr.—who began his foreign service career in 1949 and eventually became ambassador to five Arab countries (United Arab Emirates, Oman, Kuwait, Bahrain, and Qatar)—became an adviser on Arab investments. Upon his retirement in 1976, he was appointed managing director of William Sword and Company, Inc., in New Jersey, and works primarily with Arab petrodollar accounts.

Marshall W. Wiley, ambassador to Oman under the Carter administration, organizes the extensive Middle East activities of the prestigious law firm of Sidley & Austin as the firm's partner in its Washington, D.C., office. Parker Hart, former ambassador to Saudi Arabia, is a consultant to the Bechtel Corporation. Michael Sterner, former ambassador to the United Arab Emirates, heads up International Relations Consultants, Inc. Many of Sterner's clients are American multinationals exporting to the Arab world.

It is not unusual for government officials, after their retirement or forced departure, to become "consultants." Indeed, after a lifetime of government service, many officials realize the only instant commodity they can offer to the private sector is their contacts and access to the areas of the world they covered. But in the oil-rich Arab world where the lines between politics and trade have been deliberately blurred, access to and contacts with the ruling elites are often granted on the basis of political gratitude. Spiro Agnew's early fulminations against American Jews, though crude and widely condemned in the United States, provided the earliest demonstration of how this informal process worked.

The new livelihoods of these former ambassadors depended on being able to demonstrate to their corporate clients that they could contact, arrange meetings, or instantly communicate with Arab banking, government, and business leaders, who often turn out to be one and the same in the Arab world. And the most successful way in which the former envoys have retained their access to the Arab elite has been to propagate the Arab point of view before the American public and policymakers—the two most prized targets in the battle for influence over American foreign policy.

The problem is not that the envoys have perpetuated their

"localitis" perspective into their new careers. Rather, the problem is the obverse: the success of their new careers depends on the maintenance of their Arabist views. And through the skillful exploitation of the media, prestigious policy organizations and other public policy platforms, these ex-envoys have succeeded in making money from their support for arms sales, excoriation of Israel and "Zionist influence," and general parroting of the Saudi, Syrian, or another Arab government's public line.

Newspapers and television news programs have unwittingly become vehicles for the political agenda of individuals seeking private, commercial gain, which by extension is linked to the interests of Arab governments. Moreover, the problem is compounded because millions of readers or viewers of these "commentators" logically assume that the interests of these "former ambassadors," as they are identified, represent the American national interest. After all, these officials constituted the American government and are presumed to still reflect the interests and values of the American government.

To the obvious delight of Arab officials, the views of the ex-ambassadors are treated with respect untainted by any popular suspicion whatsoever that they might reflect the interests of another country.[23] Indeed, the emergence of this new diplomatic petrocorporate class, who travel across the country drumming up support for the Arab states, has greatly obviated the earlier need by Arab countries to have registered agents do their political bidding in the United States.

Beyond the issue of former ambassadors exploiting their political opinions, a larger issue concerns the ever-growing "revolving door" in Washington. To be sure, the problem—whereby officials leave government to go into the private sector and vice versa—has been around in Washington for a long time. But the involvement of so many former U.S. officials, who were entrusted with varying degrees of control over American Middle East policy, with petrodollars has created a "revolving door" of monumental and dizzying proportions. With the knowledge that their future incomes could well depend on engendering the liking of Arab governments, American officials might very well factor decisions accordingly while in government. Weapons sales, human rights assessments, tax policies, Arab investment and acquisitions, energy policies, and hundreds of day-to-day decisions are susceptible to this influence.

Aside from the growing number of retired diplomats (including over a hundred below the rank of ambassador or chargé d'affaires), the revolving door petrodollar community now extends well into the thousands, covering the entire spectrum of former American government officials: Treasury Department attachés, Commerce Department analysts, CIA station chiefs, senators, Farm Credit administrators, U.S. Army colonels (and many other military officers who have taken early retirement after working with Saudi Arabia for several years), customs officials, and even secretaries of state.

Here are just a few examples from the last several years: Colonel Robert Lilac, a high-ranking official and military specialist who served on the National Security Council during the Reagan administration—and who had worked closely with the Saudis, especially during the 1981 AWACs campaign, as well as with the Israeli government—resigned in early 1984 to work directly for Prince Bandar, the Saudi ambassador to the United States. (A spokesman for the National Security Council said that ''no policy or guidance'' existed for NSC staff members regarding employment with foreign governments.) George Ball, undersecretary of state in the Johnson administration, participated in ''missionary expeditions'' to the Saudi Arabian Monetary Agency, according to a senior banking colleague at the investment banking firm of Lehman Brothers Kuhn Loeb, where Ball had been a partner.

Edmund Muskie, former U.S. senator from Maine who was appointed to fill Secretary of State Cyrus Vance's prematurely vacant slot in the last year of the Carter administration, joined the law firm of Chadbourne Parke Whiteside & Wolff, at $250,000 a year. After a year and a half in the firm, Muskie made the following comment in an interview with *Parade* magazine about his new job: "I've made two trips to the Mideast. The adviser of the president of the [United Arab] Emirates is a new, close friend."[24] No wonder. Muskie's law firm's clients include numerous Arab businesses, so many businesses that in 1979 the firm opened an office in the United Arab Emirates, headed by a former senior official of the State Department's Agency for International Development. In 1982, the law firm joined forces with a Saudi law firm in Riyadh.

William E. Colby, the former director of the CIA, is senior adviser to a "political risk" Washington consulting firm— International Business-Government Counselors, Inc.—which advises American and European multinationals on political stability developments in foreign countries. "Analyses" of Saudi Arabia and other Arab countries are major component of the consulting firm's work, reflecting the needs of its corporate clients. But the consulting firm, cognizant of the need to maintain good relations with the Arab world, hosts an annual off-the-record meeting at which the head of the Arab League, Clovis Maksoud, is given the opportunity and platform to advance his organization's political interests before an audience of top corporate executives. According to one participant in Maksoud's seminars, held at New York City's Harvard Club, the Arab League official vehemently denounced Israeli "aggression" and "American support for Israel." The link between American foreign policy and increased trade was communicated so strongly that one corporate representative contacted the Arab League after one seminar to see what his company could do to improve U.S.–Arab relations.

Revolving doors rotate 360 degrees and so does the problem: new political "sensitivities"—fostered by previous experience or induced by the prospect of returning to an old position—are carved into government service. Reagan Defense Secretary Caspar Weinberger's consistent resistance to closer U.S.–Israeli relations and his concomitant embrace of Saudi policies and unlimited arms sales to that country are probably not unrelated to Weinberger's previous experience with Bechtel. In other, less lofty positions the potential effect may not be so visible. This does not mean that all individuals who have worked for Saudi Arabia have been directly influenced. Yet the possibility remains. Norman Bolz, executive director of the American Saudi Business Roundtable, an organization founded by former Ambassador to Saudi Arabia John West, that brings American executives together with Saudi leaders, was appointed in February 1984 to one of the top policymaking positions at the Internal Revenue Service. As associate commissioner for policy and management, Bolz is responsible for general administration, planning audits, and advising foreign governments on taxation.

David C. Mulford lived in Saudi Arabia from 1975 to 1984 and advised the Saudi Arabia Monetary Agency on investment strategy. Mulford became deputy under secretary of the treasury for international affairs in the spring of 1984. His responsibilities include formulating government policies for all internationally related Treasury Department issues, ranging from anti-boycott penalties to monetary agreements. Philip Habib, who served as special Arab-Israeli negotiator in the Reagan administration, worked as paid consultant to Bechtel and served on the board of directors of Pacific Resources, Inc., a Hawaiian energy firm that operated a joint venture with Kuwait, during his temporary recess from his shuttle diplomacy. Donald T. Regan, secretary of the treasury in the Reagan administration, had previously served as chief executive officer of Merrill Lynch, which has operated a division which exclusively advises the Saudi Arabian Monetary Authority on investments. And Attorney General William French Smith came from Gibson Dunn and Crutcher, which has been one of the select American law firms allowed to open an office in Saudi Arabia.

The work of most members of the burgeoning petrocorporate class is divorced from politics, as several insisted in interviews with me. Indeed, their responsibilities officially revolve around commerce, investment, and legal representation for and among American multinationals, the oil-rich Arab countries, and the U.S. government. But politics and trade in the Arab world, in contrast with all other trading partners, are inseparable. As demonstrated time and time again, Arab officials prefer to deal with political favorites, expect to be rewarded with political dividends, and will generally refuse to deal with "enemies" or those perceived to be opponents—such as supporters of Israel. Consultants, lawyers, financial investment counselors, and corporations who are dependent upon the goodwill of Arab countries know very well that a highly competitive market prohibits any estrangement of their clients.

In the end, the corrosive effect of petrodollar influence extends over a long continuum. At one extreme is the visceral, overt self-interested sycophancy exhibited by the former ambassadors. At the other end of the continuum is silence—chillingly demonstrated by one law firm's muzzling of a senior Democratic party member who was not allowed to testify against the AWACs sale. The revolving door has not been the only way in which Arab governments have been assured that their views are promoted in the United States. Another route was adopted by American corporations. And Aramco has led the way.

14

THE ARAMCO PIPELINE

We saw the lack of evenhandedness as a business.
We were not alone . . . virtually every business,
everybody that ever did business in the Middle
East saw this problem of American policy being
so against the Arabs. We saw it as a possibility of
damaging American business, American interests.

Frank Jungers, former
Chairman of Aramco

Aramco, unlike most multinationals, knew how to cooperate with the host government," declared Ismael Nuwab, Aramco's Saudi head of public relations, on the occasion of Aramco's fiftieth anniversary celebration in May 1983.

Reporting on the festivities, the *Washington Post* noted that "Aramco remains probably the soundest pillar of [the U.S.–Saudi] relationship." In this article and in one written a year and a half earlier, the *Washington Post* described in glowing terms Aramco's sensitivity to the social needs of its foreign hosts, such as the company's numerous educational programs and scholarships for young Saudis throughout the kingdom, extensive technical and management training, generous loans, and free land for its Saudi employees.[1] Not surprisingly, the Saudis regard Aramco, the largest oil company in the world, with admiration and affection.

But there are other reasons. Since 1967, Aramco has been sponsoring, facilitating, and subsidizing a broad network of political and "educational" activities in the United States designed both to create an illusory image of support for the Arab position in the Arab-Israeli dispute and to weaken public, congressional, and administration support for Israel. Aramco has disbursed more than $5 million through 1984 to scores of lobbyists, academicians, educational institutions, and even groups intimately connected with the PLO. In some instances, Aramco helped create organizations from scratch, while in other cases, Aramco entered into tactical alliances with such radical countries as Libya and Iraq in order to sustain the active political operations of various groups.

Set up after the 1967 Arab-Israeli War and broadened substantially after the October 1973 War, Aramco instituted a special funding program, referred to in internal documents as ''The Promotion of Islam, Arab Culture and International Understanding.'' This program was included in the annual charitable donations budget, of which 90 percent traditionally has been given to conventional charities in Saudi Arabia and elsewhere in the Arab world. The special funds, according to company documents, were earmarked under a special category that was to ''assist programs to disseminate balanced, objective and accurate information about the situation in the Middle East which led to the displacement of Palestinians from their homeland and the occupation of Arab lands, including Jerusalem.'' For the most part, however, the political agenda of the groups that Aramco assisted reflected an unrelenting hostility for Israel, Zionism, and sometimes even Jews.

It has been generally thought that the 1973 lobbying and public relations activities by oil company officials (see Chapter 2) marked the extent of the pro-Arab activities initiated by the Aramco consortium. Neither the internal oil company records subpoenaed by the Subcommittee on Multinationals nor the testimony by oil executives revealed that the oil companies had engaged in lobbying beyond 1973. Indeed, oil executives, in statements before the subcommittee, attempted to create the impression that the 1973 activities were aberrational and that the officials had been forced to acquiesce to Saudi political demands because of the unique political events that transpired that year.[2]

Whether by design or by accident, however, the oil companies in 1974–75 managed to shield from the subcommittee evidence of their ongoing funding operation. When asked in 1982 about Aramco's political funding operations, an Aramco vice-president, James V. Knight, said that Aramco "is not and has never been involved in such a campaign." [3]

The truth, however, is that Aramco has been involved in an extensive campaign to manipulate American opinion on the Middle East. And the process by which the recipients were selected for Aramco funding was a careful and meticulous one. Each possible candidate was assiduously evaluated for: (1) adherence to the Arab point of view on the Arab-Israeli dispute; (2) effectiveness in disseminating the propaganda; (3) ability to reach the media and contact American government officials; and (4) potential for generating an impact on American public opinion. "Requests would come in from all over the world," said a key oil company official who served in Saudi Arabia for many years, "and we would automatically study them."

According to Frank Jungers, the former chairman of Aramco, the company favored groups that required low overhead and "which could fan out and on an economic basis give [Aramco] a multiplier on the whole Arab image program in the U.S." [4] The decisions about whom to choose and how much to give were made by Aramco's fifteen- to twenty-member "donations committee" on which Aramco insiders, officials of Exxon, Mobil, Texaco, and Standard Oil, and Saudi officials served. Consultants in the United States, especially those already working for the four companies, also made recommendations and monitored the effect each recipient was generating.

After conducting a thorough review and obtaining a consensus, the donations committee would submit proposals as part of the much larger Aramco donations budget to conventional charities such as civic organizations, hospitals, and schools, primarily in Saudi Arabia, the West Bank, and other Arab countries.

The donations budget was included as part of the annual

capital budget that was sent to Aramco's board of directors. The board is composed of two directors from each of the four parent oil companies, Aramco insiders, and several Saudi officials, including Sheikh Zaki Yamani. In practice, the board would only rubber-stamp decisions that had been informally agreed on by top executive officers of the four oil companies and key Saudis prior to the board meeting. According to Aramco officials, the heads of Exxon, Texaco, SoCal, and Mobil personally approved the capital budget plans, which included the political funding program. High-level Saudi officials would occasionally make specific recommendations regarding the recipients of political donations, as well as to the desired Saudi political objective.

According to oil company sources, Aramco initially directed its attention toward groups based in Beirut, several of which were closely connected with the PLO. One such Arab group is the Institute for Palestine Studies (IPS). Set up in 1965, IPS describes itself in official literature as an "independent, non-profit Arab research organization, not affiliated with any government, political party or group." Yet it is known, as one State Department official said, as "the unofficial academic wing of the PLO." [5] With Kuwait University, the IPS publishes a scholarly quarterly journal, *The Journal of Palestine Studies,* that strives for a modicum of credibility and is widely read in the Department of State and in academic institutions.

According to records, Aramco has contributed at least $75,000 to IPS. Aramco documents reveal that the oil companies were especially attracted to IPS efforts that focused on regaining all of Jerusalem, a political issue in which the Saudis have expressed an intense interest.

Though IPS had already been solidly established prior to receiving Aramco's contributions, the same was not true for another group, Americans for Justice in the Middle East (AJME), also known to have worked with PLO strategists.

AJME was formed after the 1967 War by a handful of Americans associated with the American University of Beirut. As a result of the funds provided by Aramco, coupled with other subsidies from Arab sources and wealthy "individuals" in Lebanon, AJME was able to expand its base of political activities. This included a monthly newsletter, *AJME News,* mailed out to an estimated 15,000 American citizens and officials, and other activities such as a full-page *New York Times* advertisement that appeared nine days after Yasser Arafat delivered his famous "guns and olive-branch" speech to the United Nations General Assembly in November 1974. The advertisement called upon the United States to immediately recognize the PLO.

According to a 1981 newsletter, the AJME was concerned with "Zionism's virulent 30-year campaign of hate and vindictiveness." Issues in recent years of *AJME News* have focused on "Israel's nuclear menace," "Apartheid in Israel," and "Israel's exploitation of intellectuals and its violations of human rights." AJME also arranged interviews for journalists with the PLO. Between 1967 and 1980, Aramco contributed over $120,000 to finance the group. In approving funds for this group, oil officials took special note of the fact that AJME "sends its Newsletter to American opinion makers, sponsors lectures by prominent Arab spokesmen before American audiences, and sends cables and letters to U.S. Government leaders on American Middle East policy."

After the June 1967 War, two Lebanese groups were created to promote the Arab position. The Fifth of June Society and the Arab Women's Information Committee were set up exclusively to disseminate material to the American and European public. "The control of U.S. policy and media by the Zionists" was a common theme in their publications. Both groups, working with official Arab representatives, also assisted in providing "informational assistance" to Western visitors and journalists.

Aramco became a major contributor in 1968, providing over $70,000 during the next five years to these three organizations. Aramco was so pleased with the pro-Arab message promoted by these groups that it authorized its employees to start a chapter of the Fifth of June Society in the company's headquarters in Dhahran in 1968.

Throughout the groups' existence, Aramco was kept continuously briefed on their activities. In 1974, these groups merged to form the Lebanese Association for Information on Palestine (LAIP), and according to oil company sources, Aramco then increased its support to a minimum of $25,000 annually. A centerpiece of LAIP's activities in the United States was a public relations campaign in newspapers, such as the *Washington Post* and *New York Times,* which blamed Israel for the "death and destruction" in the Middle East and advocated that Israel be replaced by a "secular-democratic state."

Despite their funding of the radical Lebanon-based groups, oil company officials were still under pressure to ensure that American audiences were "exposed" to continued criticism of Israel. So the oil companies began to refocus their attention on various American organizations that could serve as ready-made political platforms.

The well-established Middle East Institute, located near Dupont Circle in Washington, is a haven for longtime Arabists, especially ex–foreign service officers and ambassadors serving in Arab countries. Established in the 1946, the institute is considered one of the bulwarks of the traditional Arab lobby in the United States. The institute is active in generating support among the Washington intelligentsia for Arab position by sponsoring conferences, speakers, and programs. A "research" paper on the Israeli economy issued in March 1983 concluded that aid to Israel should be cut substantially. Though the institute's funding is closely held, an official revealed in a 1975 interview with the *Washington Post* that oil company contributions provided 42 percent of the institute's 1974 budget of $408,086. But, said Malcolm Peck, the institute's secretary, "In the 4½ years I've been here, I haven't seen an attempt by the oil companies to try to influence our policies. They have given us pretty much carte blanche to program as we wish."[6] In that same interview, Peck denied that the institute had taken a pro-Arab advocacy position. "We always aim at presenting a balanced view."

Yet, according to Aramco documents, the Middle East Institute was considered a favorite recipient of the oil companies' largesse because of its pro-Arab disposition:

This highly respected institute has the general aim of spreading knowledge about the Middle East in the United States. MEI continues to publish its prestigious Middle East Journal, *to carry on speaker activities, and to maintain an extensive book and film library open to the public. MEI's annual conference in 1973 dealt with worldwide energy demands and the Middle East and featured a speech by H. E. Ahmed Zaki Yamani. MEI officers are known for their interest in developing accurate and balanced information about the Palestine problem.*

Between 1973 and 1981 Aramco provided the Middle East Institute with over $150,000. According to a confidential source familiar with the institute's funding, donations from Aramco and other multinational corporations have provided the bulk of the institute's revenues since 1974.

Another longtime organization is Amideast, whose board of directors is well represented by officials from American industry (e.g., Exxon, Continental Oil, and Westinghouse Electric). Though its annual report states that it seeks to "promote human resource development in the Middle East" largely through educational exchange programs, its exclusion of Israel from its services is not an accident. The group, originally called American Friends of the Middle East, was formed in the early 1950s to support pro-American regimes in the Middle East. Through the Dearborn Foundation, the organization, according to published reports, was the recipient of secret CIA funding during the 1950s and early 1960s. On a 1982 visit to its offices on Massachusetts Avenue in Washington I saw flyers promoting *Occupied Palestine*—a film touting the Arabs' rights to all of Israel—stacked neatly next to its annual reports. The map of the

Middle East printed in Amideast's annual report has routinely omitted Israel's name.

Following the 1982 Israeli invasion of Lebanon, the president of Amideast, Orrin Parker, wrote a letter to American members and supporters asking them to contact their representatives and senators about "Israel's massive aggression" that "has seriously harmed American interests in the Middle East and in the world generally." Aramco has contributed more than $175,000 to Amideast.

Another group that has received Aramco funding is American Near East Refugee Aid, Inc. (ANERA). Dedicated to helping Palestinian refugees, mostly in the occupied territories, ANERA has participated in community development projects, filled basic welfare needs, and organized political action against the Israeli occupiers. In the United States, however, ANERA's activities suggest that it is not concerned only with providing humanitarian assistance to Palestinian refugees. Letters, brochures, and newsletters sent by ANERA's officers routinely state that the Palestinians "were exiled from their homeland" in 1948 and that, consequently, "most of their country became the State of Israel." The reality is a bit more complex: many Palestinians were indeed kicked out by the advancing Jewish armies or by underground terrorist groups, but many Palestinians left to avoid the fighting (after Arab leaders beseeched them to do so). In addition, much of the territory designated by the United Nations for a Palestinian state was annexed by Jordan, not Israel. ANERA carefully avoids any suggestion that Arab governments might share some of the responsibility for the plight of the Palestinians. Nor does the group seem to care about the miserable conditions suffered and political repression endured by Palestinian refugees living in Arab countries; very little of ANERA's funds are spent to help the hundreds of thousands of Palestinians who live in refugee camps in Jordan, Syria, and other Arab confrontation states.[7] In fact, 85 percent of ANERA's funds are spent on the West Bank, Gaza Strip and southern Lebanon.

From 1971 through 1980, Aramco gave ANERA $183,000.

Other American companies have also provided millions of dollars to ANERA. During the middle of the October 1973 Yom Kippur War, Gulf Oil—whose president at the time was on ANERA's board of directors—gave $2.4 million to ANERA.[8]

According to IRS documents, others have contributed over $1 million collectively to ANERA between 1971 and 1978. Some of these contributors are:

Esso Middle East	$226,000
Esso Standard of Libya	50,000
Standard Oil Co. of California	81,000
Exxon Corporation	100,000
Texaco, Inc.	89,000
Mobil Oil Corp.	60,000
Mobil Foundation Inc.	45,000
Upjohn Company	174,000
Lederle Corporation	86,000
Pfizer	56,000
Eli Lilly	8,800
Ashland Oil Company	20,000
Amerada Hess Corporation	15,000
Warner Chilcott	39,000
Bristol-Myers	56,700

Another veteran organization to which Aramco has provided major financial assistance has been the Association of Arab American University Graduates (AAUG). Headquartered in Belmont, Massachusetts, the AAUG's positions, according to a profile in the *Washington Post,* are "militantly pro-Palestinian, pro-Arab and radical."[9]

In recent years AAUG has become more aggressive in reaching influential sectors of the American public. Since 1978, it has sponsored numerous trips to the Middle East for church organizations, professors, lawyers, trade unionists, and university administrators. It has also provided funding for American journalists to visit the West Bank in order to write articles for American periodicals about the Israeli occupation. In 1979, Jesse

Jackson was the keynote speaker at AAUG's annual convention. AAUG then pledged $10,000 to Jackson's organization, People United to Save Humanity (PUSH). (Jackson's group was also the recipient of direct Arab government funding. In 1979, PUSH received a $10,000 donation from the Libyan embassy in Washington. In 1981, PUSH solicited and received a $100,000 contribution from the Arab League—a grant which was disclosed by *New York Times* reporter Jeff Gerth in 1984.)

Between 1974 and 1983, Aramco was one of AAUG's principal sponsors, providing over $200,000. According to oil company documents, Aramco officers were impressed with AAUG's "information which seeks to present the Arab world in a sympathetic way to the American public," especially to American business leaders. This includes publishing papers and books, sponsoring lectures at universities and featuring Arab leaders at its annual conventions.

In the *Washington Post* article about AAUG, Milton Viorst wrote: "The A.A.U.G. would willingly accept funds from the PLO and oil producing countries to expand its activities, but he [Hisham Sharabi, a Georgetown University professor who served as head of AAUG] noted regretfully that they never offered any." In fact, between 1973 and 1980, according to confidential IRS documents and American intelligence officials, AAUG received more than 40 percent of its total donations from an Arab bank known for its PLO connection and from oil-exporting countries such as Libya and Iraq. Between 1973 and 1980, the largest donors, besides Aramco, were these:

Iraq Interests Section, Washington	$100,000
Arab Bank, Ltd., Lebanon	80,000
Libya	35,000
Arab League	31,600
Organization of Arab Petroleum Exporting Countries (OPEC)	20,000
Arab Information Center	10,000
Katy Industries	50,000

The $80,000 donation by the Arab Bank, Ltd., was given in April 1980 as one lump sum—the largest single amount ever received by AAUG. With $11 billion in assets, Arab Bank, Ltd., is one of the largest commercial banks in the Arab world. Arab Bank, Ltd., is also widely known for its close links to the PLO; the bank's head was also chairman of finance for the PLO in 1964. Currently, the bank reportedly handles the PLO's $60- to $100-million portfolio.[10]

Though Iraq broke diplomatic relations with the United States in 1967, it disbursed its money to AAUG via its Interests Section in Washington. An Iraqi press official in Washington denied knowledge of any funds, but according to American intelligence officials, the Iraqi Interests Section over the past decade has transferred funds to various groups in the United States.

One of these groups is the Institute for Arab Studies, a research organization founded by the AAUG in 1979, which shares the AAUG's three-story building in Belmont. A main function of the institute is to dispense "research" grants. In 1982, according to IRS documents, the Institute for Arab Studies gave $94,964 to "9 individuals to conduct specific research projects on the Arab world." One of those who received a grant was Alexander Cockburn, a columnist for the *Village Voice* and a bitter critic of Israel, who received $10,000 to write a book about the 1982 Israeli invasion of Lebanon. The grant was disclosed by Alan Lupo, a writer for the *Boston Phoenix*.[11]

Another grant—of $35,600—from the institute went to Richard Wirthlin, President Reagan's public opinion pollster, to conduct a public poll of American opinion on the Middle East. Results purportedly showed a shift toward support for the Arab position, but two polling experts interviewed in the *Washington Post* claimed the questions in the poll were "loaded" and "tilted."[12] A larger issue was raised by William Safire who recalled Saudi Arabia's use of Patrick Caddell, Jimmy Carter's pollster, and questioned whether Wirthlin should "hire himself out to a pro-Arab organization that would dearly like to influence" President Reagan.[13]

An official of the Institute for Arab Studies told Lupo that "most of [its funds] were from here [the United States]." [14] But IRS records reveal that the Iraqi Interests Section provided the institute's first, and largest, grant of $250,000 in 1979.

Though the established organizations were advocates of the Arab cause, and thus helpful to Aramco, the oil companies went a step further in one case: as disclosed by the *Washington Post* in 1975, Aramco created a nonprofit organization called Americans for Middle East Understanding (AMEU). [15] It has received, and continues to receive, most of its funding from Aramco and foreign governments, according to internal company documents and IRS records.

In the sixteen years of its existence, AMEU has evolved into a major organization within the Arab lobby. Its operations demonstrate how much Aramco has received as a return from its investment.

Among its principal activities, AMEU publishes a bimonthly newsletter, the *Link,* which has a regular circulation of 50,000. In some years, as many as 100,000 additional copies of the newsletter or other AMEU publications have been sent out. The recipients include clergy, libraries, universities, academics, businessmen, high school teachers, State Department officials, and members of Congress and their staffs.

In past issues, reference to Israel has been uniformly critical, focusing on charges of "human rights violations," denial of academic freedom, the power of the "Israel lobby," and a highly opinionated recounting of the "USS *Liberty* Affair"— the incident in which Israel attacked a U.S. ship during the 1967 War—seventeen years after it happened. References to the Arab world have accentuated only the positive: "Jordan Steps Forward," for example, and "Kuwait: Prosperity from a Sea of Oil." In 1981, AMEU actively promoted the sale of the advanced radar technology planes to Saudi Arabia. AMEU's book list, which it offers at subsidized prices to the public, is a highly selective one that aims to disparage Israel, Zionism, and even American Jews. According to descriptions in AMEU's publications, the list includes books which "refute the biblical claim of Zionists to the Promised Land," "how Jews have exercised their political power in the United States," and "Zionism's subversive consequences for both Jews inside and outside Israel."

During and after the October 1973 War, AMEU sent Mailgrams to college presidents urging that they "weigh very carefully all actions and words in support of Israel." The message went on to state that "many Arabs [are] convinced that Congress is under Zionist control" and then charged that "America's partisan news media and huge sums spent on vicious forms of advertising by Israeli sympathizers contribute to alienation among Arabs." In another mailing in late December 1973, to service station owners and members of the trucking industry, AMEU sent packets of brochures in support of the Arab view.

One AMEU official told me that Aramco's funding was "not enough," and another denied receiving foreign money, but IRS and oil company documents tell a different story.[16] AMEU has received $972,000 of its $1.8 million in donations raised between 1968 and 1982 from Aramco, representing 54 percent of its funding.

In addition, AMEU in recent years has also received substantial contributions from Libya, the Arab League, Kuwait, a Saudi prince, and other Arab sources. With the exception of 1969 and 1981, Aramco has given anywhere from 30 to 70 percent of AMEU's annual contributions. (In 1969, AMEU's total contributions of $97,000 came principally from a $26,500 Aramco grant and a $50,000 donation from Mrs. DeWitt Wallace of *Reader's Digest*.)

Between 1977 and 1982, according to IRS documents, contributions to AMEU in excess of $4,999 included:

Prince Khali bin Sultan Bin Abdulaziz	$20,000
Libya	13,000
Ambassador Amin Hilmy of the Arab League	8,000
Arab Information Center	
(a registered agent of the Arab League)	5,000

U.S. Arab Chamber of Commerce	5,000
Muslim World League	
(an organization funded by Saudi Arabia)	5,000
Fayez Sayegh	
(now deceased, he was a Palestinian	
nationalist and a member of Kuwait's	
foreign ministry and widely credited	
with authorizing the 1975 United Nations	
resolution equating Zionism with racism)	6,000
United Saudi Industries	6,000
Robert Marsh and Friends	
(Marsh has been a longtime consultant to	
Suliman Olayan's investment companies)	66,000
Fluor Corporation and Foundation	30,000
Marathon Oil	5,500

Aramco's eagerness to use a variety of spokesmen led to some rather curious political alliances. One involved funding to institutions known to be severe critics of multinational corporations. The National Council of Churches (NCC) is a left-of-center organization that has bitterly criticized the multinationals for maintaining investments in South Africa and for interfering in the internal affairs of other nations. Yet records show that NCC took money from Aramco to help subsidize a special Middle East newsletter, *SWASIA*. In fact, the Aramco money was specifically earmarked to expand circulation of *SWASIA*. During its four years of publication from 1974 through 1978, *SWASIA*'s editorial line was in keeping with the NCC's one-sided critical attitude toward Israel. Aramco officials justified the donation to *SWASIA* in internal oil company documents on the grounds that "it would present a balanced and factual picture on the Palestine problem." One issue of *SWASIA*, dated December 17, 1976, featured a translation of an Arab newspaper (*Al-Quds*, published in East Jerusalem) editorial that stated: "We expect the United States to stand firm against the American Jews and to demand that they be loyal in everything to the American flag first."

Another strange bedfellow has been Alfred Lilienthal, a self-described "progressive." Lilienthal, a Jew, is well known in the Arab world for his embrace of the Arab point of view and fierce criticism of the "Zionist" lobby and its influence. He has lectured throughout the country, appeared on numerous radio shows, and written several books, usually excoriating the role of "Zionists" and criticizing the American government for supporting Israel. In years past, Lilienthal has derided the popular media for giving so much attention to the Holocaust, calling it "Holocaustomania." To promote his views, and those of guest columnists, Lilienthal publishes a monthly newsletter, *Middle East Perspective,* with an estimated 7,500 circulation. In the past, he has also placed a full-page advertisement in the *Wall Street Journal* to promote his views.

According to oil company sources and documents, Lilienthal received at least $20,000 from Aramco between 1976 and 1980. The contributions have been specifically earmarked to "support and increase the circulation of [his] newsletter." When contacted, Lilienthal said he did not receive any contributions from Aramco. But he added that Aramco purchases a large number of subscriptions, which could "account for annual receipts of over $5,000 from Aramco."

On several occasions, it was Aramco's relationship with a particular individual, and not only the proven record of an organization, that led to the company's willingness to provide contributions. One such group that has received substantial funds was the American Committee for Justice in the Middle East (ACJME), a group based in Boulder, Colorado.

In the United States, ACJME's activities included publishing dozens of well-researched special reports, placing full-page advertisements in American newspapers, sponsoring conferences, and lobbying senators and congressmen. ACJME argued for a cutoff of all aid to Israel and a realignment with the Arab nations, and condemned the role of "Zionists" in American government. One ACJME report issued in 1975 painted a new picture of the 1973 oil embargo: "The [oil] embargo, at first called blackmail by the Israeli lobby, turned out to be non-military means of pressing for peace negotiations in the Middle East."

According to oil company sources and documents, ACJME secured $10,000 a year for several years in the mid-1970s. Aramco's donations committee's recommendations specifically noted ACJME's "factual testimony before the Democratic and Republican Party Platform Committees," "its letters published in the *New York Times* and *Christian Science Monitor*," and "special position papers and reports." Aramco also backed the group because of its "active protest(s) against one-sided American military assistance in the Middle East."

Professor Ragei El-Mallakh, who was active in ACJME, is also linked to another Aramco-supported program. Oil company records show substantial disbursements to the "Middle East Economic Program" at the University of Colorado, a program headed by El-Mallakh. Aramco has made annual contributions of at least $10,000 to this program during the eleven years since it was founded in 1973. And, according to University of Colorado sources and IRS records, other major contributors have included Shell, Texaco, Exxon and the East-West Foundation (funded by the Fluor Corporation).

Aramco officials, according to oil company documents, considered El-Mallakh "one of the most effective Arab spokesmen on the American scene." The oil company was especially impressed by the fact that El-Mallakh's "writings have appeared in high school text books used by an estimated 10 to 14 million students." Aramco "observers" in the United States, the oil documents note, "spoke highly of his effectiveness in presenting a factual account of the Middle East situation to American audiences." And El-Mallakh's articles, Aramco specifically pointed out, "have appeared in the *New York Times* and *Wall Street Journal*."

When a university spokesman was asked about the "Middle East Economic Program" at the University of Colorado at Boulder, he responded that no such official university program existed. However, he said that El-Mallakh did operate his own independent research center, out of his campus office. Called the International Research Center for Energy and Economic Development, it publishes the semiannual *Journal of Energy and Development*. The university did not provide funding or any other support, the spokesman added.

In fact, however, conferences run by the center take place at the university and, moreover, the inside cover of the journal states that it is issued twice yearly "under the auspices of the International Research Center for Energy and Economic Development *and* the University of Colorado" (emphasis added).

The journal is an impressive publication, with articles over the past two years from a dazzling array of international professionals, including OPEC officials, energy experts supportive of OPEC, oil company officials, and even U.S. officials such as Congressman David Stockman, before he became the head of the Office of Management and Budget in the Reagan administration. The center itself sponsors numerous illustrious conferences featuring high-ranking OPEC officials. One held in March 1982 included presentations by an aide to Sheikh Yamani and a former secretary general of OPEC. The center has provided OPEC with a prestigious (and supposedly nonpartisan) public relations outlet in the United States.

An even more surprising Aramco arrangement has transpired at the Johns Hopkins School of Advanced International Studies (SAIS). According to oil company records and sources, Aramco may have had influence over the material used in one course on the Middle East that was taught at the school for eight years. Oil company documents show that Aramco contributed over $120,000 between 1973 and 1980 to cover the costs of offering one course, The Arabian Peninsula: Government and Politics. Aramco officials, according to internal oil company records, were pleased with the contents of the course both for its "accurate information" and "concentration on Saudi Arabia" and attached much importance to their influence with such a prestigious university: "SAIS has special significance in being located in Washington, and in having many government leaders attending its courses."

Between 1973 and 1980, the course was taught by Professor John Duke Anthony, who is no longer there. It is now taught by another professor, and an examination of university records has revealed that the curriculum has been entirely revamped from that used by Anthony. According to sources close to SAIS, Aramco lobbied specifically to retain Anthony at SAIS when his rotating nontenured position expired in 1980–81. Anthony is now a consultant to, and lecturer for, several prominent pro-Arab lobbying groups in Washington. In addition, in 1984 he became the head of the National Council on U.S.-Arab Relations, a newly formed "educational" organization substantially composed of former U.S. ambassadors to the Arab world, which dispenses grants and holds symposia around the country in support of pro-Arab understanding.

A spokeswoman for the university conceded that Aramco, like many other corporate and noncorporate donors, has made substantial contributions to SAIS. "But," said public relations official Susan Crowley, "no donor is allowed to make any strings-attached donations. No contributions can affect courses. That is university policy." She said that the receipts from Aramco showed designations for "the Middle East Program," which covers many courses, or for "general purpose" use.[17]

Why did Aramco agree to serve as a conduit of funds? Frank Jungers, former Aramco chairman, explained that the oil companies had taken political actions and funded pro-Arab groups because "it amounted to the price of doing business in the Kingdom."[18] In 1977 the fifty-seven-year-old Jungers retired as head of Aramco. He now serves as a business consultant, primarily to the Bechtel Corporation, and lives in Portland, Oregon.

The special donations program, Jungers said, was the product of "being told what to be by the Saudis" and of actions "taken to please the Saudis." Jungers said that Aramco had only one interest in mind, "to retain the [oil] concession." He continued, "When somebody like the King would call me in and express his dismay or inability to do what he thought we ought to do, we had to respond to this request in some way or at least show that we were interested in their aspirations." And, he added, "we simply couldn't fail to cooperate on a reasonable request." Undermining popular support for Israel, a foreign policy obsession of Saudi Arabia, was, according to Jungers, considered legitimate and proper by Aramco. He saw nothing wrong with funding groups linked to the PLO or with making surreptitious arrangements with American universities that resulted in tampering with academic curricula.

Was it proper, I asked, for Aramco to play such a role? "What do you do?" he responded rhetorically. "The image of the oil companies, which was tied to the Arabs by the press of this country, was not a good one. And in my opinion, that image was not justified. So what do you do? Well, you simply set out to try and set the record straight. And there were many programs that Aramco did along these lines, over a long number of years."

The development of Jungers's strongly held views and his ultimate involvement in explosive political matters attests to the political conversion process many Americans underwent after working in Saudi Arabia for a period of time. As a twenty-one-year-old graduate in 1947 with a degree in mechanical engineering from the University of Washington, Jungers signed up with Aramco but planned to stay at most two years. He soon realized that Aramco, starved for young American professionals, presented unusual growth opportunity.

The five-foot nine-inch native of Regent, North Dakota, was dispatched from Aramco's offices in Saudi Arabia to Beirut, Lebanon, for one full year in 1962 to learn Arabic. His rise to the top of the consortium began in 1968 when he was appointed Aramco's top official for government relations and liaison with the

Saudis. A year later he was promoted to vice-president and entrusted with the responsibility of negotiating with Arab governments on behalf of Aramco. By 1971, he was made president and two years later, chairman.

The soft-spoken Jungers made no attempt to ascribe Aramco's activities exclusively to Saudi pressure; rather, he is a sincere advocate for the Arab cause, a position that indisputably evolved through years of wearing Aramco spectacles: "We saw the lack of evenhandedness as a business. We were not alone; virtually every business, everybody that ever did business in the Middle East saw this problem of American policy being so against the Arabs. We saw it as a possibility of damaging American business, damaging American interests. . . . See, you had a press at that time that was violently anti-Arab, and the question that had to cross any American's mind who had anything to do with the Middle East was: Is this really good for the country?"

Jungers shed some additional light on why Aramco became involved with radical groups with extensive connections to the PLO. Aramco needed to placate the authorities and political groups in the countries through which the 2,000-mile Trans-Arabian Pipeline (Tapline) ran. The steel pipeline passes through Jordan, Syria, and Lebanon, where tankers unload their oil for shipment to Europe. "Aramco had to contribute to those [Beirut-based] groups," Jungers asserted. "You bet. We did it as part of our business. . . . It was part of the necessity of doing business in the [Tapline] countries."

I asked Jungers why Aramco wasn't willing to disclose its millions of dollars of contributions to political groups, yet publicly touted its glossy, expensive, and nonpolitical *Aramco World* magazine? Didn't this project a disingenuous image of noninvolvement in the Arab-Israeli dispute? Jungers responded, "We weren't looking for an argument. We were looking for both sides to be seen. What was happening in the press was totally Israeli." Jungers candidly admitted that public disclosure by Aramco of its funding program "would have discredited the Arab cause. And that wasn't for us to do."

He continued, "What we had to do in the long run—really what the Saudis had to do—was to get general public understanding. It's as simple as that. . . . And then, at the same time, if questioned by the Saudis—what are you doing?—we could point [to these programs]. We're not sitting by and living in this country and doing absolutely nothing. To me, it's a question of enlightened self-interest."

Other oil company executives I interviewed also expressed strong views on the Middle East; each one invariably echoed the official Saudi line.

The Aramco campaign, the first of its kind by an American firm, soon led to a succession of other corporate-funded programs designed to engender goodwill toward Saudi Arabia among the American people. But the political conversion process—induced by the simple prospect of garnering trade with the Arab oil producers—had made it unnecessary for Saudi and other Arab officials to press their American partners for specific action: the Americans had been transformed into true believers. Or, if they didn't believe, they certainly knew the rules of the game they so desperately wanted to play. Promoting the Saudi gospel through shrewd manipulation of the media, universities, public television, books, lectures, and other public opinion channels would soon be the order of the day. Academia was next.

15

ACADEMIA
FOR
SALE

I don't see why the PLO has to have
a PR organization when Georgetown
is doing all their work for them.

Art Buchwald

Halfway between Columbus and Cincinnati sits the small Ohio town of Wilmington, where an 800-student Quaker school is located. In late 1982, Wilmington College officials distributed one-page flyers on campus announcing a forthcoming "Convocations on the Middle East." Ten outside speakers would deliver lectures on subjects ranging from Islamic art to contemporary Middle East politics. The series was described on the flyer as part of the school's "continuing effort to provide increased understanding of international issues." What students did not know when they read the circular was that the entire lecture series had been organized and funded by the American Educational Trust (AET), a Washington-based "educational" organization headed by former ambassador to Qatar Andrew Killgore and funded substantially by American businesses and Arab donors. The petrodollar connection had come to Wilmington.

Established in 1982, AET began its first year, according to IRS records, with $1,072,237 on hand—an unusually large amount for a new organization.

The prospectus of AET states that it is "one of many institutions seeking to make Arabs and Americans aware of the mutuality of Arab and American interests." But the principal way in which the organization has found to promote this "mutuality of Arab and American interests" has been to focus on the evils of "Zionism" and the state of Israel.[1]

At Wilmington, a group of faculty—aware of the not-too-terribly secret partisan nature of the AET—protested the absence of any mention of the organization on the campus circular. A new brochure was soon disseminated by the college that identified the AET as the sponsor of the lecture series. But the college's description—"the American Educational Trust [is] one of many institutions seeking to make Arabs and Americans aware of the mutuality of Arab and American interests"—had been lifted straight from the organization's prospectus.

Between January and March 1983, ten "experts," handpicked and paid by the AET, were flown to Wilmington. When a group of Wilmington professors suggested that an additional speaker be incorporated to balance the views of a particular lecturer—a former government official known for his ardent embrace of the Arab point of view—the faculty members were told by the college official in charge of the program that AET would not allow tampering with the format. The request was denied.

Wilmington was not the only beneficiary of AET's assistance. The group routinely provides speakers at no charge—like Killgore's presentation to the University of Indiana—to scores of colleges and universities. And another Washington organization devoted to "reinforc[ing] the mutually beneficial ties between the United States and the Arab nations" also dispatches political speakers to colleges: The American Arab Affairs Council is funded, according to a council source, by large donations from American multinationals and wealthy "individuals and businessmen in the Persian Gulf." Among the firms that have contributed to the Council or subsidized its publications (through advertisements) are Fluor, Northrop, Hughes Aircraft, and the Boeing Commercial Airplane Company. The American Arab

Affairs Council was founded in 1981 by former American foreign service officers who served in Arab countries. The council's current president is George A. Naifeh, who had previously worked for the United States Information Agency in the United Arab Emirates, Oman, Jordan, Libya, Algeria, and Pakistan. The secretary and treasurer is Roderick M. Hills, a partner in the law firm of Latham Watkins and Hills. Hills had once been a presidential counsel in the Ford administration and was also a chairman of Sears Roebuck and Company. The American Arab Affairs Council's diplomatic advisory committee consists of eleven former American ambassadors: Lucius D. Battle, Egypt; Herman F. Eilts, Saudi Arabia and Egypt; Parker Hart, Saudi Arabia; Andrew Killgore, Qatar; Wilbert J. LeMelle, Kenya and Seychelles; E. Allan Lightner, Jr., Libya; Donald F. McHenry, U.S. Representative to United Nations; Talcott Seelye, Syria and Tunisia; Michael Sterner, United Arab Emirates; William A. Stoltzfus, Kuwait; and Marshall W. Wiley, Oman.

The council's most successful "outreach" activity has been its special conferences revolving around a cleverly fused theme: the interlocking of American economic and political interests in the Arab world.

Presentations on exporting to the Arab world and the dimensions of Arab investment are routinely mixed with speeches denouncing Israel and the Jewish lobby by leading Arab officials and former U.S. officials, most of whom are members of the petrocorporate class. Occasionally, a de facto competition of sorts has erupted among the speakers to see who can portray the Jewish lobby in the most sinister manner. Ironically, the Arab officials have been consistently less intemperate in their speeches than their American counterparts.[2]

But what is most unusual is that the American Arab Affairs Council has succeeded in getting the official sponsorship and financial support of major corporations and universities for these highly political conferences. At one conference, for example, held in St. Louis, Missouri, in September 1983, McDonnell Douglas Corporation and General Dynamics Corporation, both huge defense manufacturers, were among the corporate sponsors, along with St. Louis University and the World Affairs Council of St. Louis.[3] For a conference at the University of Wisconsin at Milwaukee in November 1983, Allis-Chalmers and the First Wisconsin National Bank of Milwaukee provided financial support.

The activities of these Washington-based groups are overshadowed, however, by the proliferation of direct "strings-attached" grants by Arab governments to major universities across the United States. The Saudis, who led the way, made their contributions to carefully selected recipients who guaranteed a good return on their investment.

In fact, by 1978, over ninety American colleges and universities had tried to obtain direct Saudi aid.[4] Of the ninety schools that sought such aid, only three were initially successful in their quest: the University of Southern California, Duke University in North Carolina, and Georgetown University in Washington, D.C. The three were strategically located in different parts of the country.

Of these three, the first recipient was the University of Southern California (USC). In 1976, it received a Saudi grant of $1 million to establish the King Faisal Chair of Islamic and Arab Studies. Though the endowment of chairs in American universities is hardly rare, USC granted the Saudi government an extraordinary privilege: the right to approve the selection of the chair's occupant and participate in the selection of all of its future occupants. Professor Willard A. Beling, a professor of international relations who had previously worked for Aramco, was appointed to the chair less than five weeks after a letter was sent on March 26 from a Saudi government official to USC President John Hubbard stating: "It is our understanding . . . [that] the first incumbent of the Chair shall be Professor Willard A. Beling."

Saudi Arabia's choice of USC as the first American university to receive such financial largesse was not mere happenstance. USC had long been a favorite of many Saudi officials. An Old Boy Network had sprung up, due to the scores of illustrious Saudi graduates, including Dr. Gazi Algosaibi, the Saudi minister of industry and electricity; Dr. Soliman Sulaim, the minister of commerce; Hisham Nazer, minister of planning; fourteen deputy ministers; two hundred Saudi businessmen, academicians, and other government officials.[5]

In 1977, an alumni chapter of the university had even been started in Saudi Arabia at a dinner party attended by USC President John Hubbard. Hubbard had become a believer in the USC-Saudi connection. Interviewed in 1978, Hubbard—whose office was adorned with a picture of him and Saudi King Khalid—said, "I am absolutely convinced that they've been moderate on oil policy in OPEC because of the USC connection."[6]

In 1978, the relationship between Saudi Arabia and USC took a quantum leap forward thanks to the actions of J. Robert Fluor, the chairman of USC's board of trustees. Fluor was the head of the Fluor Corporation, a California firm started by his grandfather in 1912, which had emerged as one of the world's largest engineering and construction firms. The Irvine-based firm built multibillion-dollar mega-projects such as oil refineries and oil-drilling platforms throughout the world, ranging from China to South Africa. Saudi Arabia was Fluor's biggest customer, generating more than $5 billion in contracts from the construction of the Saudi gas-gathering system. The fifty-six-year-old Fluor, a former U.S. Air Force pilot who raced thoroughbred horses, had become an early Saudi supporter among the American business community. In 1975 his firm produced favorable documentary-style films on the kingdom for showing throughout the United States. And in March 1978, Fluor himself sent a letter to 40,000 of his company's employees, stockholders, and vendors, as well as to members of Congress, urging them to support the sale of F-15 air superiority planes to Saudi Arabia.

In May 1978—the month the Senate voted on the F-15 sale—Fluor endeavored to prove his Saudi loyalty on a much more ambitious scale. He invited forty executives from the top American multinationals—such as Mobil, Litton Industries, and Exxon—to a breakfast meeting at the Biltmore Hotel in Santa

Barbara, California. His purpose was to raise funds for an elaborate $22-million Middle East Center designed to enhance understanding of Saudi Arabia. The center was to operate autonomously within the University of Southern California "to prepare both undergraduate and graduate students for academic, business and governmental careers relating to the Middle East; to facilitate academic research on the area; [and] to provide research and related services concerning the Middle East to the nonacademic community."[7]

The Middle East Center was also to be given a voice in the selection of faculty in other departments of USC. The fact that the proposal still awaited official university approval was not mentioned to the executives. Citing the need for good public relations, two prominent guest speakers, Saudi Foreign Minister Prince Saud al-Faisal and Industry and Electricity Minister Algosaibi—"whose approval," the New York Times noted, "is required for any major industrial deal with the Saudi government"—made strong pitches for contributions. The Saudi Arabian government, however, would not make any donations, said Algosaibi, because that would undermine the credibility of the center.[8]

Fluor immediately followed the Saudi speakers with a direct appeal to the businessmen's self-interest. "Contribute," he told them, "and your company will be remembered by Saudi Arabia."[9] As the executives departed from the hotel, they were handed a brochure describing the Middle East Center plus individually tailored requests for contributions ranging up to $1 million.

By September of that year, over $7 million had been pledged. The money was supplied by a handful of American firms whose executives had attended the breakfast with Fluor, and from the East-West Foundation, a nonprofit charitable arm of the Fluor Corporation. On October 4, 1978, the trustees of the university voted to approve the center despite the fact that they would not have total control over the center's decisions and program, such as curricula and selection of faculty. Many faculty, however, protested the arrangement and actively challenged the trustees' decision. In a unanimous vote, the faculty senate voted to disapprove of the arrangements and procedures as academically unacceptable. Yet, the trustees would not budge from their support of the center.

At the same time Fluor was involved in another controversial episode. The East-West Foundation, which claimed to be independent of Fluor but in fact received 65 percent of its funds from Fluor and was run by a public relations consultant for Fluor, provided a $650,000 annually renewable grant to the Aspen Institute, an independent academic and research organization based in Colorado. The money was to pay for an "Islamic-Middle East" program that would, among other things, "focus international attention on regional developments and problems." It soon became clear that the focus of this new program would be through the Arab perspective.

Characterized by academic integrity, the Aspen Institute had developed an international reputation for its well-balanced public policy conferences drawing experts from all over the world. So, in preparation for a seminar on "The Shaping of the Arab World" facilitated by the East-West Foundation grant, two Israeli scholars were invited by Aspen officials along with Arab and European scholars. Suddenly, only weeks before the scheduled start of the July 1979 seminar, the Israelis were told by Aspen officials that their presence would be unwelcome. When asked about this episode, the president of the Aspen Institute, Joseph Slator, said, "[We] are not the United Nations. We didn't feel the need for an Israeli quota." [10]

The demands imposed on Aspen by the Fluor-supported East-West Foundation were unprecedented for a contributor. Other Aspen officials told the *New York Times* that an East-West Foundation official, Christopher Beirn, who was also a consultant for Fluor, repeatedly threatened to terminate all contributions to Aspen unless Israelis were barred from participating in several conferences and unless a specific program was moved from the venue of Aspen's choice, Jerusalem. [11] When the controversy exploded in the media Aspen officials beat a hasty retreat and decided to stop accepting funds from the East-West Foundation.

A similar fate occurred to the Fluor-supported USC Middle East Center. After a lengthy debate, which received extensive and mostly critical news coverage, the USC trustees voted unanimously in June 1979 to reverse their original decision and nullify plans for the proposed Middle East Center.

Twenty-eight hundred miles to the East, in Durham, North Carolina, another jointly sponsored Saudi-American corporate program fared much better. This time efforts had been taken to ensure that the program did not become too overtly politicized, at least in the beginning. Since its inception in 1977, the Program in Islamic and Arabian Development Studies at Duke has flourished as the only academic center in the United States officially devoted to the study of the "Arabian Peninsula"—in other words, Saudi Arabia. Yet, in spite of the absence of shrill rhetoric that has characterized other Arab studies programs, the program at Duke has emerged less along the lines of a typical academic program and more like a de facto southern branch of the Saudi embassy.

According to an internal Duke memorandum, the founders of the program—and the Saudi benefactors—expected the program to provide for the "diffusion of information" throughout the "eastern part of the United States," specifically to create a regional balance to the USC program.[12]

Two hundred thousand dollars in seed money was provided by the Saudi government in 1977, thanks to the "vigorous support," the Duke memorandum stated, of three Saudi ministers: Algosaibi, Soliman A. Sulaim (both of whom had been involved earlier in the USC program), and Mohammed Abdu Yamani, minister of information.

Three years later the grant was doubled to $400,000.

In addition to its academic offerings—ranging from classes on Arabic literature to a course stressing the significance of Jerusalem to the Islamic world—the program's activities include:

hosting lavish conferences bringing together Saudi officials and American potentates; offering prestigious public relations platforms to Saudi officials during their visits to the United States; providing consulting services to American firms doing business in the Middle East; and sending out guest lecturers to numerous southern colleges to explain the Saudi perspective on the politics of the Middle East.

The Arab studies program, an annual Duke report states matter-of-factly, has become "a clearinghouse of information on Islam, the Arab world, and Saudi Arabia in particular," specifically noting that requests have come in from CBS, the Department of State, local newspapers, and magazines across the country.

On the Duke campus the program has drawn criticism from some faculty. Commented Arif Dirlik, professor of history and chairman of Duke's East Asian Studies Committee, "It is less a university activity than an activity in which certain members of the university serve as a go-between for Arab interests and major corporations." In response, the director-founder of the program, Ralph Braibanti, said, "That's not true. It's a scholarly activity. We have a very impressive publication record that speaks for itself." [13]

Unlike most financially strapped academic programs, the Duke program has had no problems raising funds. Saudi Arabia provides 60 percent of the program's annual expenses; the remaining funds come in from a host of multibillion-dollar corporations and their foundations such as Bechtel, Exxon, Mobil, Triangle International, Northrup, Lockheed, Standard Oil of California, J. A. Jones Construction Company, Daniel International, East-West Foundation (Fluor), and Aramco (one of its few publicly acknowledged contributions). Arab government organizations and businesses also contribute.

In September 1979, the program hosted a major conference that brought together the largest number of Saudi officials and scholars ever assembled for an academic conference. Organized with input of Saudi Information Minister Yamani—who inaugurated the conference—seventy Saudi and American officials, businessmen, and academicians, in addition to reporters from the *New York Times, Wall Street Journal,* and *Christian Science Monitor,* converged at Duke. At the three-day affair all aspects of Saudi Arabia were discussed, ranging from manpower problems to archaeology and politics.

Fouad al-Farsy, Saudi deputy minister of industry and electricity—and the first Saudi to get his Ph.D. at Duke—was the lead speaker on one panel devoted to foreign affairs and security. He spoke about the "external threats" to Saudi Arabia posed by Iran and the peace treaty between Egypt and Israel. His speech was followed by a talk given by a Saudi registered agent, William Fulbright.

Though politics was not the official theme, the political undercurrents running throughout the conference, according to several participants, were palpable, particularly regarding Israel. The prevailing attitude seemed to have been summed up in the last paragraph of an editorial in the *Middle East Economic Digest* written by Joseph P. Malone—one of the conference's participants and the head of a Middle East consulting firm that promoted exports to the Arab world—and reprinted in the Duke program's annual report. Malone expressed his appreciation to both U.N. Ambassador Andrew Young—who had just met illicitly with the PLO representative and was later fired as a result—and to Jesse Jackson—who, immediately following Young's resignation, flew to the Middle East where he hugged Yasser Arafat. Malone said, "Given the need for economic growth and for Saudi Arabia to be an oil producer for another 60–80 years—at 8.5 million to 9.5 million barrels a day—one must agree with the comment of one U.S. participant, 'We must be thankful to Andrew Young and to Jesse Jackson for giving freedom of speech to so many of us.' " [14]

On hand to greet the participants was Duke's president Terry Sanford, a longtime southern proponent of civil rights, who expressed his belief and hope that the "conference was very much a position of Duke University's concern for what we might learn from other parts of the world."[15] Sanford became sensitized to the riches afforded by the chief sponsor of his university's new program. Six months later, Sanford and Braibanti, the director-founder of the program—who was appointed in 1984 to serve on the policy planning staff at the State Department—traveled to Saudi Arabia at Saudi invitation where they met with dozens of top Saudi education and government officials. Later that year in October, Sanford hosted a dinner in honor of a "visitor" to the program, Saudi businessman Ghaith R. Pharoan.

The program's greatest impact has been achieved through an ambitious outreach program in the southern United States. The founders of the program intended for it to ultimately reach "25,000 students in three key Southern states" and generate an impact on "journalists, publicists, and communications media in all eight states of the Southeastern region." In addition, "the circulation of thirty distinguished outside lecturers" was expected to reach "academic and non-academic audiences of approximately 40,000 persons."[16] There is every bit of evidence that the Duke Arab studies program is well on its way toward reaching its goals.

During 1978, the program's first year in full-time operation, Duke arranged to send outside lecturers—whose expenses and honoraria were paid by the League of Arab States—to a group of small liberal arts colleges in surrounding states. The lecturers ranged from Islamic art experts to highly partisan pro-Arab political commentators such as John Duke Anthony and other specialists with pro-Saudi dispositions.

In the years since, the outreach program has substantially expanded. Professors from Duke have been dispatched to nearby colleges to lecture on "contemporary events and security concerns in the Arabian Peninsula"—and thanks to special, hefty grants from Texaco and the Exxon Educational Foundation, additional speakers have been brought in, without cost, to a dozen local colleges.

In late 1981 Exxon began to provide funds directly to the affiliated colleges to expand or create miniature Islamic and

Arabian studies programs à la Duke. After sponsoring a $50,000 summer workshop at Duke for visiting professors in 1981, officials of the Exxon Educational Foundation sent letters to the presidents of the twelve southern colleges that had been represented at the summer program. The twelve recipients were the College of Charleston, Berry College, Belmont Abbey College, Appalachian State, Coker College, Converse College, Davidson College, Livingstone College, Presbyterian College, Salem College, Johnson C. Smith University, and St. Andrews Presbyterian College. Enclosed was a $100 check for "phoning and postage" so that the faculty member who attended the summer session at Duke could keep in touch with Duke officials, and something more tantalizing: a promise of an immediate $1,000 to help set up new courses and lectures on Arabian and Islamic studies for the campus and community at large. All twelve of the colleges responded affirmatively to the offer. They have since created programs and scheduled speakers, all of which have been specifically approved and funded by Exxon Educational Foundation officials in New York.

At Converse College, a small women's college located in Spartanburg, South Carolina, for example, the Exxon grant facilitated the acquisition of special films, books, and lectures designed "to expand knowledge of Arab culture" for students and townfolk. The expansion of "culture" on Converse's campus has taken a heavily pro-Saudi coloring as evidenced by the speakers brought in from Washington: John Duke Anthony and Joseph P. Malone.

Though Exxon officials whom I interviewed denied that the program has any political bias, Joe Dunn, professor of history and politics, who administers the Exxon grant at Converse College stated otherwise: "Clearly the purpose of the whole Exxon program is to have an Arab point of view. In addition, I feel personally that the Israeli point of view has had more than fair play." He freely acknowledged that he received $1,000 to set up a program to "expand knowledge about the Middle East from the Arab perspective."[17]

In the nation's capital, another elaborate program was established with Saudi money. Unlike the other two, it received funds from numerous Arab governments. In the spring of 1975, Georgetown University officials visited a handful of Arab officials seeking underwriting for their proposal to organize a Center for Contemporary Arab Studies. Although Georgetown already had a healthy Arabic department that functioned as part of the college, the proposed center was to operate autonomously; it was to serve as the country's first full-fledged quasi-independent academic institution focusing exclusively on Arab affairs.

After meeting with Sultan Qabus Bin Said of Oman, Dean Peter Krogh received a $100,000 check. Krogh, head of Georgetown's School of Foreign Service, had taken the lead in helping to set up the new center. The Sultan tripled his commitment of $100,000 in the next four years. During that same time, other Arab grants came from Egypt ($145,000), Jordan ($15,000), Saudi Arabia ($200,000), Qatar ($70,000), United Arab Emirates ($350,000), and Arab ambassadors in Washington ($55,000).[18]

After just two years, the center at Georgetown was able to expand substantially, offering a full range of courses such as Arab history, politics, sociology, philosophy, language, and religion. However, it wasn't long before the center assumed an air of partisanship that seemed to overstep the bounds of academic objectivity and neutrality.

After the Israelis invaded southern Lebanon in 1978 following a PLO attack on a Tel Aviv bus, Clovis Maksoud, a visiting lecturer as well as a chief spokesman of the Arab League, gave a briefing for Washington reporters in which he bitterly condemned the Israeli attack. Though his briefing to the Washington journalists was not unusual, the manner in which the university treated his comments certainly was: Georgetown University published his words in an official university press release. Afterwards, columnist Art Buchwald wrote in the *Washington Post*, "I don't see why the PLO has to have a PR organization when Georgetown is doing all their work for them."

A year before, in May 1977, Buchwald had been critical of Georgetown's acceptance of a $750,000 donation from Libya for endowment of the Umar al-Mulkhtar Chair of Arab Culture. In a letter to the student newspaper, Buchwald charged the university with taking "blood money from one of the most notorious regimes in the world today." He also asked whether the university might also set up a "Brezhnev Studies Program in Human Rights or an Idi Amin Chair in Genocide." Georgetown's faculty at the Center for Contemporary Arab Studies disagreed with Buchwald's criticism. Said Dean Krogh: "I don't know Uganda. I've never been to Uganda. I don't know Idi Amin." The center's director, Michael Hudson, also responded to Buchwald: "The Libyans say they are just as anti-terrorist as anyone else."[19]

But Georgetown University's president, Father Timothy S. Healy, decided in February 1981 that the Libyan money had to be returned. After dropping off a check for $642,721 at the Libyan embassy, he announced that "Libya's continued accent on violence as a normal method of international policy and its growing support of terrorism as a tool of government has made it increasingly impossible for Georgetown to feel comfortable in having its name associated with the Libyan Government." Hisham Sharabi, the activist professor—an unabashed supporter of the PLO and former head of the National Association of Arab Americans—who had been appointed to occupy the Libyan chair, disagreed: "The Libyans are very decent, very thoughtful, very considerate, and very correct" in endowing the chair.[20]

This was not the first time that Georgetown regretted having solicited and accepted money. In the summer of 1978, the school returned a $50,000 donation to Iraq—the money was to fund research in Islamic ethics—following a public outcry. For a university to return funding, especially an amount of this magnitude, is almost unheard of in the perennially impoverished world of academia. Father Healy, to his credit and to the applause of many editorial writers—and also to the biting criticism of Sharabi who called Healy a "Jesuit Zionist"—rejected the Libyan and Iraqi money.[21] His decision even resulted in a surprise contribution of $100,000 to the university by the investment house of Bear Stearns in appreciation of his action.

But somehow lost among the uproar over the Libyan donation was a far more serious fund-raising episode that raised the question of whether Arab countries were able to directly purchase political goodwill. In January 1980, Krogh, Hudson, Sharabi, and retired vice-admiral Marmaduke G. Bayne—who formerly commanded the U.S. Middle East Force in the Persian Gulf—traveled to five Persian Gulf countries in search of additional funds. In their sixteen days abroad, they met with more than fifty government, educational, and business leaders.

Immediately upon their return to Georgetown, the four embarked on a public relations offensive, speaking out strongly against American policy in the Middle East. Hudson urged the United States to "support the PLO" in order to "preserve [the] status quo." Krogh, according to the *Hoya,* the undergraduate newspaper, "called for a reconciliation of America with Arab public opinion." And Sharabi declared that "the perceived threat [to the Arabs] is not . . . the Soviet Union [or] communism . . . it is Israel and Zionism."[22] Then the group wrote and distributed a special report to various members of Congress, the media, and officials at the State Department. The report relayed "Arab views" on "major issues" and offered a set of policy "recommendations." Virtually the entire report blasted Israeli policies ("a breach of international law and civilized behavior"); condemned the Israeli government (a "theocratic state"); and criticized current American policy (the United States "must immediately and visibly demonstrate its cultural respect for and human interest in the Arab world"). Several days later, on February 6, portions of the report were inserted into the *Congressional Record* by Congressman Paul Findley, Republican of Illinois.

In mid-March, Krogh sent letters to several congressmen urging them to support an amendment, a copy of which he enclosed, to cut aid to Israel by $150 million, the sum, he claimed,

Israel was spending on "Jewish settlements in the occupied territories." The amendment, he added, "was in keeping with the recommendations of a report with which I was associated."

Within the next nine months, a staggering $2.75 million in Arab government money flowed into the center. The donations came from the United Arab Emirates ($750,000), Kuwait ($1,000,000), and Oman ($1,000,000)—all of which the Georgetown delegation had visited in January. The two $1-million gifts represented the largest foreign gifts ever received by Georgetown.

Adding to Georgetown's newfound riches were hundreds of thousands of dollars contributed by scores of American businesses with an eye toward pleasing Arab governments. According to university records, these included: American Broadcasting Companies, Inc., Bechtel, Chase Manhattan Bank, Ford Motor Company, General Electric, General Motors Corporation, Getty Oil, Morgan Guaranty Trust Co., Mobil Oil, Rockwell International, Otis Elevator, Texaco, United Technologies, and Whittaker Corporation.

As a result of the influx of millions of dollars, Georgetown's Center for Contemporary Arab Studies has become one of the largest Arab studies programs in the United States. The true price, though, has been a political one: the Arab world's obsession with Israel and American identification with Israel have also become the obsession of the Georgetown program. Virulent criticism of Israel and American support for Israel are the single most dominant themes of the center's extremely active program.

Many of the symposia, films, lectures, publications, and colloquia are more directed toward generating political sympathy and support for the Arab cause than toward creating a legitimate and open academic environment.[23] At the Model League of Arab Studies, a program supposedly designed to familiarize students with the workings of the League of Arab States, the proceedings have replicated the real world with a frightening reality: various resolutions—some written with the help of the center's professors—were approved at one Model League proceeding that condemned "Zionist" influences in government and media and the "racist ideology" of Israel. In short, the center— which, boasts a university report, "is used as an important media resource at the local, national and international levels"— functions like an advocate and public relations organization rather than a neutral academic program.

Elsewhere across the country Arab donors made other efforts to attach strings to their donations. At the campuses of Swarthmore, Haverford, and Bryn Mawr, three highly regarded liberal arts colleges in Pennsylvania, a Swarthmore administrator circulated plans for a proposed $590,000 "Arab Studies" program in early 1977. The aims of the program were spelled out in unusually frank terms: "understanding and sympathy for the Arab point of view" and "encouragement of a favorable [Arab] public relations climate in this country." [24]

Despite the transparent political motives of the proposal, the three colleges jumped at the opportunity to participate—only to withdraw after they found out that the sponsor of the program was the foundation of Adnan Khashoggi. Khashoggi at that time was just emerging as a principal character in the multinational payoff scandal involving American defense manufacturers.

To the State University of New York at New Paltz and to the University of Pennsylvania, Libya offered $180,000 in late 1977 for the development of new high school course material on the Middle East. Several officials and faculty of these schools accepted expense-paid trips to Tripoli. New Paltz, however, soon rejected the Libyan program. The University of Pennsylvania tentatively accepted the funding. But in the wake of unsatisfactorily answered questions by faculty about whether the research would be objective and whether Jews might suffer discrimination, the university rejected the Libyan proposal.

In the summer of 1983, secret negotiations were held between representatives of the State University of New York at Stony Brook and high-ranking Saudi officials to create an $11-million Islamic studies program. [25] According to internal university documents, contacts were made with Prince Sultan bin Fahd,

King Fahd's son; former Saudi Ambassador Faisal Alhegelan; Aramco; the King Faisal Foundation; and several corporations. The document noted that "Litton [Industries] with a $4 billion Saudi contract would be happy to contribute in terms and unit multiples of $100,000 to, say, $500,000." If the program is created as proposed, it will constitute the largest Saudi-supported Islamic studies center in the United States.

Universities were not the only vehicles by which Arab governments and their American corporate allies were able to spread the Arab message. And as demonstrated by the efforts of Fluor, sometimes American companies took the lead in orchestrating public relations. Bechtel produced a series of films and even commissioned an author to write a book, *The New Arabians*, which was distributed by Doubleday. The book glorified the history of the Arabian Peninsula.

PBS broadcast a three-part series on Saudi Arabia in the guise of a "documentary." Saudi Arabia itself could not have commanded a more effective and favorable means of transmitting its views to millions of Americans on "Zionism," "AWACs," the political favors it has done for the United States, and even Saudi "doubts about the U.S. as an ally."

Written, produced, and narrated by Jo Franklin-Trout, former producer of the much-acclaimed "The MacNeil-Lehrer Report," the series featured prominent "experts" such as John West and James E. Akins—identified only as "former Ambassadors to Saudi Arabia." The lavish, expensive production, shot mostly on location in Saudia Arabia, was funded by four American corporations, none of which were oil companies.

To the average viewer, and to the millions of high school students across the United States who received specially prepared guides on the series and on Saudi Arabia, the documentary's credentials must have seemed impeccable.

What few people knew was that the documentary originated in part from the efforts of the State Department to placate Saudi anger following the broadcast of *Death of a Princess* in

1980. From the very beginning, the "documentary" promised to be a whitewash when covering political issues.

Each of the four companies kicked in $140,000—two of them sponsoring a PBS program for the first time—for reasons other than tax write-offs. Morgan Guaranty Trust of New York is one of the largest repositories of Saudi funds in the United States, handling billions of the kingdom's petrodollar investments. The second donor, Texas Instruments Incorporated, owns a little-known but very important company, GSI, which has operated in Saudi Arabia since the mid 1930s and was responsible for discovering much of Saudi oil reserves. The third donor, the Harris Corporation, sells tens of millions of dollars' worth of telecommunications equipment to Saudi Arabia. And the fourth benefactor, Ford Motor Company, which has been on the Arab blacklist for years, has been openly looking for ways to engender Saudi goodwill. In fact, in 1982, Ford supplied 80 million dollars' worth of technology and equipment to a consortium of twenty-one Arab governments and the PLO that were building a vast satellite communications network called Arabsat.

Neither the corporations, the State Department, nor the most important participant—the Saudis—were disappointed with the final result.

Neither were the Saudis disappointed in the decision of a major American cultural institution, the Smithsonian, to cancel an exhibition of Israeli artifacts. In January 1984 Smithsonian officials abruptly announced that the long-awaited exhibit, Archaeology of Israel, would not be shown. Kennedy Schmertz, director of the Smithsonian's Office of International Activities, explained that the exhibit had been canceled because the "ownership [of 11 items out of a total of 320] was disputed."[26] Those eleven items had come from the Rockefeller Museum, located in the Arab sector of Jerusalem, which Israel had captured in 1967.[27] According to Smithsonian sources, the curators for the impending show were shocked by the decision to ban the exhibition, on which they had worked for eighteen months.

Indeed, in the spring of 1982, the Smithsonian had no qualms about those eleven items: at that time, it gallantly agreed to sponsor the exhibition after the Metropolitan Museum in New York had rejected the Israeli show.[28] But something happened in the intervening two years. In December 1983 the Smithsonian announced that Saudi Arabia had agreed to donate $5 million to build a Center for Islamic Studies. The contribution was the largest foreign gift the Smithsonian ever received.

Though Saudi and Arab oil influence was generated throughout the United States as a result of strings-attached donations, another way in which the petrodollar impact was clearly felt was through investment in the American economy. Nowhere was the effect more visible than in the extraordinary efforts of the Ford, Carter, and Reagan administrations to keep secret from Congress and the public vital information about Arab investment in the United States.

16

THE SECRECY PLEDGE AND ARAB INVESTMENT

[D]isclosure of this information would
be likely to cause grave injury to our
foreign relations.

*February 17, 1982, letter
from President Reagan to
Congressman Rosenthal
on why he was vetoing
the release of information
about Arab investment
in the United States.*

An hour before the scheduled start of the 10:00 A.M. meeting of the House Subcommittee on Commerce, Consumer and Monetary Affairs on Thursday, May 6, 1982, a dozen CIA security personnel arrived at the congressional hearing room. Once inside the second-floor room in the Rayburn House Office Building, the CIA contingent divided into two groups: one group "swept" the room for electronic transmission devices; the other stood guard outside the chamber, despite the fact that only the Capitol Police Force has jurisdiction over the Capitol buildings. By the time the hearing started, ten more CIA officials had joined their colleagues.

During the two-and-a-half-hour closed-door hearing, an official House reporter recorded the proceedings, including the testimony of various CIA officals, for preparation as a transcript. (Prior to the hearing, the CIA sought to prevent the making of any transcript.)

At the conclusion of the hearing, a CIA security official took possession of the stenographic tapes and, together with seven other CIA officials, escorted the House reporter to the subcommittee offices three floors below to pick up additional materials for the transcript. The CIA's plans were to take the tapes back to the Agency's headquarters in Langley, Virginia, where they were to be "stored" and "transcribed" by CIA personnel.

When Peter Barash, the subcommittee staff director, and Theodore Jacobs and Stephen McSpadden, the two subcommittee counsels, glanced out of their offices, they were startled to see the House reporter surrounded by CIA security personnel. Barash inquired what was going on. The CIA congressional liaison told him that the CIA was going to take the tapes back to Langley. Barash was astounded. The tapes were the property of Congress, not the CIA. "This was not a meeting of the Central Intelligence Agency," he exclaimed. "This was a meeting of a Subcommittee of the Congress!"

A shouting match ensued between Barash and the CIA liaison. A standoff continued until the chairman of the subcommittee, Congressman Benjamin Rosenthal, was reached on the floor of the House. He instructed his staff not to allow the CIA to take possession of the tapes. The Office of the Clerk of the House was also contacted; a House lawyer reaffirmed Rosenthal's decision. The tapes belonged to Congress, he said, and were not to leave the Capitol. The CIA reluctantly obliged.

What was the subject of the testimony that led the CIA to impose such stringent security and to attempt to take control over the operations of a congressional subcommittee? What had induced the CIA to go to such lengths to prevent disclosure of its testimony when its briefings to Congress on international terrorism and covert operations do not receive such electronic or physical protection? Why had President Ronald Reagan become personally involved two months before the hearing in an effort to stop the congressional subcommittee from attempting to declassify materials that were discussed at the hearing?

The answers to these questions revolve around an area that has elicited a policy of unusual government sensitivity: Arab investment in the United States. The CIA's actions on May 6, 1982, represented only one of many fantastic efforts taken over a ten-year period to keep Arab investment data permanently shrouded in secrecy.

Since 1974, the Ford, Carter, and Reagan administrations have refused to divulge to the public, press, and U.S. Congress critical information about the extent and nature of investment in the United States by Saudi Arabia, Kuwait, and the United Arab Emirates. In fact, the executive branch has suppressed all data concerning the individual level of investment by Saudi Arabia, Kuwait, and the United Arab Emirates. And congressional committees attempting to obtain pertinent information about such investments have encountered fierce resistance by the executive branch.

At a cursory glance, the officially recorded amount of Arab investment in the United States would not seem to warrant the secrecy exhibited by the executive branch. By the end of 1983, according to Treasury Department statistics, the level of investments by "Middle East oil exporters"—Treasury will not provide a country-by-country breakdown—in the United States stood at $74.6 billion.[1] This group includes Bahrain, Iran, Kuwait, Oman, Qatar, Saudi Arabia, and the United Arab Emirates. Of the $74.6 billion held by these countries, $39.9 billion—or 53 percent—was in the form of U.S. government securities, such as Treasury bills and bonds. The remaining investment was divided as follows: $5.1 billion in corporate bonds; $8.6 billion in corporate stocks; $6.7 billion in deposits in American banks; $4.3 billion in non-bank liabilities (such as private loans to American corporations); $5.1 billion in debts from the U.S. government (such as prepayments for military purchases); and $4.8 billion in "direct investment." A direct investment is the purchase of 10 percent or more of an American corporation.

Deposits in offshore branches of American banks are tallied separately. As of December 31, 1982—the last and apparently final year for which the Treasury Department has released these types of statistics—the deposits of Middle East oil-producing countries in the offshore branches was $13.3 billion.[2] Of these deposits, 74 percent—or $9.8 billion—was concentrated in just six American multinational banks: Bank of America, Chase Manhattan, Chemical Bank, Citibank, Manufacturers Hanover, and Morgan Guaranty. These Arab deposits, however, represented only 2.9 percent of the total deposits of $342 billion held in these six banks.

The same relatively small percentage characterizes the investment position of "Middle East oil exporters" when compared to other foreign investment in the United States. Investments from Japan, Germany, and the United Kingdom far exceed Middle East investments. Of the $781.5 billion total foreign investment in the United States at the end of 1983, the investments of the "Middle East oil exporters" constituted 9.6 percent. Though Middle East investment declined by $8 billion from 1982 (largely as a result of a reserve draw-down by Arab governments to compensate for lower oil revenues), its percentage of the total foreign investment in the United States has never exceeded 13 percent, according to Treasury Department figures.

Yet the political leverage of Arab investors over the actions of three presidential administrations has been inversely disproportionate to these holdings.

In testimony before Congress, Treasury Department officials—who maintain primary responsibility over foreign investment data—have persistently maintained that the confidentiality accorded to Middle East oil countries is required by law. These officials have justified their policy of denying Congress a country-by-country breakdown of Arab investment by invoking the International Investment Survey Act of 1976 and the Bretton Woods Agreement Act, two congressional statutes covering the collection and dissemination of foreign investment data by the executive branch. Both acts prohibit any public disclosure of investment information relating to "individual investors." The Treasury Department has asserted that the primary investors of the Arab oil-producing countries are the governments themselves and their central banks, thus entitling Saudi Arabia, Kuwait, and the United Arab Emirates to the confidentiality afforded "individual investors."[3] In effect, as pointed out in a report by the General Accounting Office (GAO), "Treasury's decision appears to be one of protecting individual human beings [but] [i]n fact they are withholding information concerning massive transactions of official Government monetary institutions."[4]

Though Treasury's rationale supposedly provides the basis for not releasing a country-by-country breakdown to Congress, it does not explain why Treasury has refused to share Arab investment data with other federal agencies, especially since the two statutes specifically authorize the sharing of investment data. The Commerce Department, for example, complained in an internal 1978 memorandum that Treasury's refusal to supply Arab investment data "hinders" the Commerce Department's ability to monitor international transactions and to "uncover" statistical deficiencies in the country's balance-of-payments figures.[5] Moreover, Treasury will not allow the governors of the Federal Reserve to see individual OPEC country breakdowns.[6] And during the Iranian hostage crisis, Treasury would not provide Iranian investment data to an emergency ad hoc interagency. This group of consultants—chaired by Under Secretary of State David Newsom and composed of representatives of seven federal agencies, including the CIA, National Security Council, and Defense Department—"needed the data to assess the consequences of a possible Iranian withdrawal of assets from the United States."[7]

The real reason why Treasury has assiduously protected the confidentiality of Arab investment data has little, if anything, to do with legality or legislation. Rather, the policy stems from demands and ultimatums by Saudi Arabia and Kuwait, whose representatives made it clear that if they were not accorded special privileges, they would withdraw their investments or reduce oil production.

In December 1974, when Arab oil producers began accumulating tens of billions of dollars more than they could spend, even after making massive expenditures for their huge modernization programs, the Arab governments needed to invest their money. And in order to get the Saudis to invest in the United States, the American government acceded to the request of the Saudi government to suppress any disclosure of its investment position. But not until more than four years later did Congress first learn of this unprecedented agreement. In March 1979, auditors for the Government Accounting Office (GAO), the investigating arm of Congress, interviewed William Simon, who was secretary of the treasury at the time the Arab governments started to amass their huge wealth.

To induce the Saudis to purchase large blocks of U.S. securities, GAO reported, Simon cut a deal with the Saudis. GAO's auditors reported on their interview with the former Treasury secretary: "In exchange for these security purchases, the United States assured the Saudis confidentiality in reporting data on them by region. According to Mr. Simon, 'This regional reporting was the only way in which Saudi Arabia would agree to the deal for add-ons.' "[8] The "add-ons" were a new facility specifically created by Treasury to accommodate the large Saudi purchases of Treasury securities. Prior to Treasury's announcements of regular offerings of securities, the Saudi Arabia Monetary Agency (SAMA) was to be contacted to see whether it was interested in purchasing additional amounts of the same issue. If the Saudis wanted these securities, they would be allowed to purchase them at a prearranged price, without having to bid in a public auction, where American investors would later purchase these same securities.

Treasury has also asserted repeatedly in testimony before Congress that it has provided to the Congress and to the public an adequate amount of information necessary to "address and determine the implications of OPEC investments for U.S. national interests." In fact, the executive branch's own track record raises serious concern: the Treasury Department and other federal agencies have proven unable to detect or monitor all of the Arab investments in the United States. Through offshore tax havens (such as the Netherlands Antilles), dummy corporations, laundered money, Swiss bank accounts, and the lack of adequate government controls, many billions more of Arab surplus funds have entered the United States than are officially reported or known. In fact, one Arab leader actually revealed his intention to establish conduits for certain types of investments as another Arab government has done. According to a State Department cable, a top leader of a "Mid-East OPEC government" (either Saudi Arabia, Kuwait, or the United Arab Emirates) told an American diplomat in June 1979 that his government would "probably establish offshore companies in the Bahamas, etc., similar to those set up by another Mid-East government, through which it would conduct real estate investments."[9]

Estimates of the amount of undetected Arab investment have ranged up to $200 billion—more than twice the amount officially recorded by the Treasury Department. David Mizrahi, the editor of *Mid East Report,* an authoritative newsletter on Middle East trade, investments, and political developments, testified before Congress in September 1981 that Arab officials told him privately that Arab investment in the United States was $150 to $200 billion.

The most damaging revelations come from the government itself. In a speech at New York University in March 1980, the chairman of the Federal Reserve, Paul Volker, stated there was $75 billion of OPEC investment in the United States that had not been picked up in Treasury statistics.

A year earlier, when the Carter administration was confronted with the Iranian hostage crisis, the State Department attempted to compile an inventory of Iranian-owned assets in the United States. According to a confidential State Department memorandum dated February 2, 1979, State Department officials reported as follows: "There is very little information available about Iranian-owned assets in the United States, since most transactions can be made without encountering any reporting requirement. . . . No one has any firm estimate of the size, distribution or composition of Iranian private real estate in the United States."[10]

As of September 1982, the Treasury Department had recorded a cumulative discrepancy of $67 billion in OPEC's investable surplus—meaning that the Treasury knows the money exists but has no idea where. At times, the inadequacy of government accounting of capital in and out of the United States has provoked incredulity from senior government officials. An internal Treasury Department memorandum prepared for Secretary William Miller in March 1980 noted a "statistical discrepancy" of $28.5 billion of foreign money in the United States in 1979. C. Fred Bergsten, assistant secretary of the treasury for international affairs, added his own handwritten comments in the memo's left margin: "This is ridiculous. Do we have any idea what is causing this! Can any more research help?" [11]

Understandably, there has been cause for serious concern in Congress. The beginning of congressional attempts to examine the problem goes back to 1974–75, when huge surplus oil revenues started to flow into the United States. By the end of 1976, Saudi Arabia and the United Arab Emirates had accumulated financial assets of over $100 billion. Like any prudent investor, these countries placed their surplus funds in the most secure and stable financial markets they could find—the economic institutions of the industrialized West.

Prominent multinational banks, like Chase Manhattan, Citibank, and Morgan Guaranty suddenly found themselves awash in petrodollar deposits. The banks were obviously delighted with new riches, which they loaned to poorer countries to help finance their exorbitant energy bills. But no one in the American executive branch focused on the long-term political implications of growing Arab oil revenue surpluses.

So, as part of its legislative supervision, the Senate Subcommittee on Multinational Corporations, chaired by Frank Church, began to look into the relationship between the burgeoning international debt, the growing concentration of petrodollar surplus revenues in American banks, and the pressures on American foreign policy.

On April 17, 1975, the subcommittee sent a questionnaire to thirty-six major banks asking for a breakdown of deposits from twenty-two countries, including the OPEC nations. The banks, however, for proprietary reasons and because they were fearful of antagonizing their Arab customers, refused to comply. Their fears were grounded in more than just speculation. In early September 1975, Kuwait's minister of finance, Abdul Rahman Atiqi, warned Senator Charles Percy and Assistant Treasury Secretaries Chester Cooper and Gerald Parsky that "Kuwait would definitely pull its funds out of U.S. banks if its position was revealed as demanded by the Subcommittee."[12]

After arduous negotiations, a compromise was finally reached that allowed the subcommittee access to some of the banking data. Two years later, the subcommittee released an exhaustive study entitled "International Debt, the Banks and U.S. Foreign Policy." Written by subcommittee staffer Karin Lissakers, the report presented a shocking portrait of the potential foreign leverage wielded over the banks and the American government.

The report warned of the emergence of a "money weapon" owing to the $50 billion in assets in the United States held by the oil producers: "At least half of these assets are highly liquid, such as Treasury bills and short-term bank deposits which could quickly be withdrawn or converted if the need arose. Any sudden movement of this volume of funds could be extremely disruptive of the financial system."

The report concluded:

> In the event of another major outbreak of hostilities in the Middle East, in which the United States and Saudi Arabia are likely to find themselves on opposite sides, can one be sure that they will continue to act in the best interests of the Western financial system? Saudi Arabia did not hesitate to use the oil weapon against the United States in the last Mid-East war, despite earlier warm U.S.–Saudi relations: There is no guarantee that next time they won't wield the money weapon, too.

The Carter administration ignored the report. In Congress, however, concern continued to mount. In preparation for a set of hearings, Congressman James Scheuer, a New York Democrat, wrote Secretary of State Cyrus Vance on May 24, 1978. In his letter, Scheuer posed "eleven very specific questions on petrodollars held by the Saudi Arabians and the potential impact on our foreign policymaking apparatus." Seven weeks later, Scheuer received a response from one of Vance's aides at the State Department. Only one of the eleven questions had been answered.

Subsequently, Scheuer learned that a detailed response to all his questions had in fact been prepared by the CIA at the request of the State Department. But the Treasury Department intervened and "excised most of the CIA draft response." [13]

Months later, the Treasury Department succeeded in imposing a de facto gag order on the CIA. [14] Following several incidents in which CIA analysts had apparently talked to the press and prepared reports for Congress about Arab investment, Secretary of the Treasury Michael Blumenthal sent a strongly worded letter on November 15, 1978, to CIA Director Stansfield Turner:

> There have been several recent incidents which have led us to believe that the extreme sensitivity of such information, particularly data on the location and currency composition of the official assets of foreign governments, may not be fully appreciated by CIA personnel. . . . I would also be grateful if the CIA would not disseminate to the Congress, on a classified basis or otherwise, information on such transactions, even if obtained solely from agency sources, without our concurrence.

Turner agreed to the request. In his response to Blumenthal, the CIA Director wrote: "I have instructed OER [Office of Economic Research, the CIA division that analyzes Arab investment data] to consult with your department before providing substantive responses—either written or oral—to congressional requests on OPEC foreign asset questions.'' But in his letter, Turner acknowledged that his decision was clearly in conflict with President Carter's Executive Order 12035, which instructed the CIA to "facilitate the use of national foreign intelligence products by the Congress in a secure manner."

The following year, 1979, Congressman Benjamin Rosenthal, Democrat of New York, began to focus on OPEC investments in the United States as part of a larger investigation by his Subcommittee on Commerce, Consumer and Monetary Affairs into the operations of federal agencies in monitoring foreign investment. In April 1979, Rosenthal sent a detailed letter to Treasury asking for relevant documents such as understandings between the United States and OPEC countries regarding the treatment of their investments.

Treasury produced 300 documents, most of which covered the period 1974 through 1977, but refused to turn over 350 other documents that were more timely. Besides invoking the two acts relating to collection of foreign investment data, Treasury officials also asserted the right to withhold sensitive material from Congress if they felt there was not adequate enough protection and reserved for themselves the right to determine which documents could or could not be disclosed by the subcommittee.

The subcommittee maintained—and was supported by legal opinions of both the Library of Congress and the General Accounting Office—that the Treasury Department's stated rationale was without merit and was in violation of Congress's right to access to information in pursuance of its constitutional responsibilities. The two acts invoked by the Treasury Department prohibited disclosure of investment data only to the public. There was no stipulation in the acts that Congress would be prohibited access to this information.

A major confrontation was brewing between the Treasury Department and Congressman Rosenthal. Attempts to reconcile differences failed.[15] After several days of hearings in which administration officials refused to turn over Arab investment data,

Rosenthal called a meeting of his subcommittee to consider the issuance of a subpoena to force the secretary of the Treasury to produce the withheld documents. The meeting was scheduled for 9:00 A.M. on July 24, 1979. But earlier that morning, Rosenthal received a surprise guest, Under Secretary of Treasury for Monetary Affairs, Anthony Solomon. "I have seen the cables," Solomon told Rosenthal, "the Saudis are very upset about the hearings. Our relations with them will be seriously affected." Solomon also told Rosenthal that the Saudis might feel "provoked" into "disposing of some of their investments."

Saudi Arabia was not alone in issuing warnings to the American government in anticipation of Rosenthal's investigation and hearings. In early July, a senior Kuwaiti official relayed to American embassy officials in Kuwait a message from a conversation he had with the head of Kuwait, Shiekh Jaber al-Ahmad al-Sabah.[16] According to a cable sent by the American embassy in Kuwait, American diplomats were told that the "Rosenthal hearings and attendant brouhaha might lead [it] to move some of its [Kuwait's] funds from the U.S. into the Eurodollar market." The Kuwaiti official also revealed that "in the past few weeks, European bankers and others had been flocking [to Kuwait] urging [the] government to put more money in European countries and Eurodollar accounts." An American embassy officer promised the Kuwaiti official that the Treasury Department would "challenge GAO assertions that U.S. government agencies are not justified in withholding data from Congress." In their cable to Washington, embassy officers acknowledged that the Kuwaiti official "seemed more concerned by [the] publicity [that the] hearings would generate than [by the] actual risk of disclosure."

Congressman Rosenthal continued to press his investigation. At the July 24 meeting the subcommittee authorized by a vote of five to zero the issuance of a subpoena to Treasury Secretary Blumenthal. But in a conciliatory move, the subcommittee elected not to execute the subpoena for the additional Treasury documents pending another round of negotiations with the Treasury Department.

Rosenthal's staffer and specialist on OPEC investment, Stephen McSpadden, a thirty-one-year-old antitrust attorney formerly with the Justice Department, headed up the investigation and also conducted the negotiations with Treasury. Agreement was reached on half of the disputed documents; but an accord on the other half proved elusive. Rosenthal reactivated the subpoena process by scheduling a hearing for early October. Meanwhile, the Carter White House, feeling added political heat from the Saudis and Kuwaitis, attempted to kill the subpoena. Robert Lipschutz, counsel to the President, became involved in trying to deter the subcommittee from taking any action. But Rosenthal successfully counterlobbied; and the prospect of an imminent subpoena induced Treasury to compromise.[17] Treasury agreed to declassify some documents, and the subcommittee agreed to accept Treasury's deletions on other documents and Treasury's insistence that over thirty-five classified documents never be released.

Negotiations handled by McSpadden now began with the State Department and the CIA for other OPEC-related documents. At one meeting held at the State Department in January 1980, McSpadden had to deal with nine State Department officials, each representing a different Middle East country and each arguing against disclosure. At one point a State Department official declared, ''We have to protect the banks' interests''—specifically mentioning Chase Manhattan. The grueling five-hour session produced no consensus. In fact, it was not until July 1981 that an arrangement was finally agreed upon.

The negotiations with the CIA also proceeded at a snail's pace. An agreement had been reached between Rosenthal and CIA Director Stansfield Turner on July 12, 1979 which set the guidelines for any future release of CIA-provided documents. Specifically, if the CIA and the subcommittee failed to resolve their differences over which documents could be released, the subcommittee would be allowed to publish the documents, with deletions protecting intelligence sources, unless the President intervened in the interests of "national security." Only a congressional resolution could override the President's decision.

Of the one hundred documents provided to the subcommittee by the CIA, McSpadden and Rosenthal identified seventeen "important but non-sensitive" documents that they felt ought to be declassified. And because these documents revealed a whole new set of problems and issues associated with OPEC investment—including changes in Arab investment strategies and increased threats to the stability of American financial markets—Rosenthal wanted to get them into the public domain. These documents included "OPEC: Official Foreign Assets"; "Kuwait and Saudi Arabia: Facing Limits on U.S. Equity Purchases"; "Problems with Growing Arab Wealth"; and "Kuwaiti Investment in the United States."

But the CIA was adamant in its refusal to allow any declassification. The matter was deemed of such importance that CIA Director Casey and his assistant, Admiral Bobby Inman, had both become involved in the dispute with Rosenthal. The CIA refused to negotiate with the subcommittee, even boycotting a hearing to which they had been invited to explain their position.

In the meantime, according to government sources, the Saudi embassy in Washington continued to make strong representations to the Reagan administration cautioning against any disclosure of its investments.

Because no resolution of the dispute with the CIA was in sight, Rosenthal wrote President Reagan on February 9, 1982, of his intention to publish the documents with appropriate deletions "to protect CIA intelligence sources." In his letter, Rosenthal explained his reasons:

> *Some of the CIA documents raised concerns about*
> *OPEC investment not expressed by other government*

agencies such as State and Treasury. In congres-
sional hearings and elsewhere these agencies have
repeatedly asserted that there is no basis for concern
in the recycling of petrodollars and OPEC invest-
ments in the United States. The CIA documents ex-
press a different view on the type of investment OPEC
governments have been making in the recent past.
Without the CIA documents, the public will get a dis-
torted and one-sided Executive Branch view of the
nature, extent and impact of such investment.

Eight days later, President Reagan responded to Rosenthal. In his February 17 letter to the congressman, the President wrote that "disclosure of this information would be likely to cause grave injury to our foreign relations or would compromise sources and methods of intelligence gathering." The President added that "the public interest in avoiding such injury outweighs any public interest served by disclosure."

Though Reagan had vetoed release of the documents, Rosenthal would not give up. There was only one avenue left. He introduced a resolution in the House to overrule Reagan's decision. But before a showdown on that resolution could occur, Rosenthal offered the CIA one more opportunity to work out a compromise. On the morning of May 6, 1982, the subcommittee met behind closed doors with the CIA officials. This was the meeting in which the CIA swept the room for bugs, posted its own guards outside the door, and tried to oversee production of the transcript. At times the session became quite heated. For example, a CIA witness maintained that an analysis of the effects of the oil glut on Abu Dhabi was too "sensitive" to be released and contended that such disclosure could compromise CIA sources. Rosenthal shot back, "You could get this from *Fortune* magazine!"

Upon the departure of the CIA officials, the subcommittee met to decide upon a course of action. After intense wrangling, the subcommittee agreed to a compromise with the CIA. It would only publish "summaries" of the documents. The vote on the compromise was six to five, with all the Democrats, including Rosenthal, voting for it and all the Republicans voting against the resolution—that is, against release of any material.

Despite the stonewalling and fierce resistance that he encountered over the course of more than four years, Rosenthal was still able to get into the public domain hundreds of documents. Even with heavy censoring and deletions, these documents, together with other published accounts provide a portrait of the tightly held world of Arab investment and the nature of Arab threats against the United States.

One batch of documents revolves around a 1978 episode in which the IRS proposed to revise and update its sixty-year-old policy that exempted foreign governments from paying any income tax on their investments in the United States.[18] Specifically, the IRS proposed to take away the exemption for foreign investments of a clearly commercial nature. While foreign government investments in stocks, bonds, and domestic securities were to remain exempt from tax, the new IRS regulations would have taken away this tax-exempt status from certain real estate investments such as "shopping centers and commercial buildings which are leased on a conventional and net lease basis."

On being informed of the proposed IRS changes, Kuwaiti officials exploded. Of all the OPEC investors, Kuwait owned the largest share of commercial U.S. real estate. Numerous meetings were held between Kuwaiti officials and American government officials in Kuwait, New York, and Washington. In one meeting in late 1978, a high-ranking Kuwaiti warned the American embassy in Kuwait that the proposed IRS revisions would "have consequences" on the level of his country's oil production.[19] Drawing on a legal memorandum written up on behalf of his country by the New York law firm of Milbank Tweed Hadley & McCloy, the Kuwaiti insisted that the IRS had been misguided in its proposed revisions. His message about future oil production was unambiguous: "I am not looking at the matter from the financial side, but from the oil side. The only incentive we have to produce is that we can use the proceeds to invest—to exchange one asset for another."

In other meetings with American diplomats, Kuwaiti officials heatedly insisted that Treasury Secretary Simon and Assistant Secretary Gerald Parsky had earlier promised in 1975 and 1976 that Kuwaiti real estate investments in the United States would never be subjected to tax. When apprised of this alleged agreement, IRS officials searched government records but found "no evidence that such commitment was ever made" and also concluded that it was "highly unlikely that it would have been made." [20]

Other government documents reveal that Kuwaiti investors—primarily the Kuwait government and prominent Kuwaiti families—have adopted an agressive investment strategy in the United States. Kuwait has acquired control of the $2.5-billion oil and drilling company Sante Fe Industries. Other investments have included the Hawaiian Independent Refinery, Inc.; the Petra Capital Corporation (investment banking) in New York; the Andover Oil Company in Oklahoma; AZL Resources, Inc. (oil refining) in Arizona; Solid State Technology, Inc. (fire alarm boxes) in Massachusetts; and Burkyarns, Inc. (yarn spinning mill) in North Carolina. Kuwait also has numerous holdings in hotels, shopping centers and office property, such as the Atlanta Hilton in Georgia; the Columbia Plaza Office Building in Washington, D.C.; the Baltimore Hilton in Maryland; the Landmark Hotel in Nevada; the Hotel Statler in New York; the Galleria in Texas; and other real estate holdings in over thirty states.

In 1981, Dan Dorfman of the *Chicago Tribune* published confidential banking documents from Citibank showing that Kuwait had acquired up to 4 percent of the stock of 197 leading American corporations. [21] Worth some $7 billion, the Kuwaiti portfolio managed by Citibank included large blocks of shares in such companies as McDonald's, Ralston Purina, Atlantic

Richfield, Johnson & Johnson, General Motors, General Mills, Dow Chemical, Eastman Kodak, J. C. Penney, and Procter & Gamble.

Saudi Arabia has followed a much more cautious investment strategy, though in the last five years it has started to change its policies. Initially Saudi Arabia invested most of its funds in U.S. government securities. In fact, by 1978, Saudi Arabia was the largest holder of Federal National Mortgage Association (Fanny Mae) certificates which, *Business Week* pointed out, "means that a lot of U.S. houseowners are directly indebted to the Arabs for their mortgage money."[22] Saudi Arabia also owns the largest share of the $39.9 billion held at the end of 1983 by Middle East oil exporters in U.S. government securities.

Saudi Arabia has become a major investor in American corporations. In fact, according to one CIA document, the level of Saudi as well as Kuwaiti investments was so high by the middle of 1977, that both countries were "having difficulties placing funds in the U.S. stock market without triggering the SEC 5 percent disclosure rules." Those rules require that any purchase above 5 percent of a corporation be disclosed publicly; investors who control less than 5 percent can remain anonymous. Major Saudi investments in the United States have included: the investment banking firms of Smith Barney Harris Upham & Co. and Donaldson Lufkin & Jenrette; the security brokerage firm ACLI International; and such banks as the Main Bank of Houston, the National Bank of Georgia, and the Bank of the Commonwealth of Missouri. Other Saudi investments in the United States include Coastal and Offshore Plants Systems (aluminum smelting) in North Carolina, Salt Lake International Center in Utah, Sunshine Mining Co. in Texas, RLC Corporation (trucking) in Delaware, Hyatt International Corporation (hotels) in Illinois, Colorado Land and Cattle Company (beef cattle) in Arizona, Plaza of the Americas (hotel) in Texas, and real estate in over two dozen states.

Aside from the political influence that such investment may create, the larger policy issue with which Congress has tried to deal revolves around the question of whether Arab investors could or would withdraw their investments to advance political objectives.

Treasury officials have insisted to Congress that Arab governments would not risk jeopardizing the value of their investments by pulling them out precipitously. And if Arab governments did pull out their funds, their argument continues, they would have to place them in other Western financial markets and would soon find their way back into the American financial system. But as revealed by the Rosenthal subcommittee's documents, the executive branch has consistently taken seriously the threats by Arab oil producers to withdraw their funds or curtail oil production if their demands for secrecy and confidentiality were not met.

Moreover, as noted by the 1977 Church subcommittee report: "even the threat of such action by the Arabs could put great political pressure on the United States, whose banks hold most of petrodollar deposits, and which has a major foreign policy stake in the Middle East."

In September 1981, the Iraqi oil minister publicly urged other Arab countries to withdraw their deposits from American banks because of an American policy that "openly antagonizes the Arabs." [23] And according to a former senior official in the Reagan administration, during the Israeli invasion of Lebanon in June, July, and August 1982, the White House received several reports—transmitted through European diplomats and other intermediaries—that unless the United States restrained Israel, King Fahd was ready to start pulling Saudi deposits from American banks and begin liquidating holdings in Treasury bills. None of these alleged threats was received directly by Reagan administration officials.

Yet, even in the absence of independent confirmation, the threat was taken seriously. Because of the magnitude of Saudi holdings and the realization that Saudi Arabia *might* use its wealth as a political weapon, the nature of Saudi leverage over the United States is such that Saudi leaders need not issue any direct threats:

the fear of any withdrawal of petrodollar investments is sufficient to instill concern.

Unannounced changes in Saudi investment policies have led to intense crystal-ball speculation as to the motivations of the Saudi decisions. Invariably, the political factor always looms.

For example, in mid-1979 Treasury statistics revealed that by March 1979, Saudi Arabia holdings in Treasury securities had suddenly fallen to $9.8 billion from a level of $12.8 billion at the end of 1977. The drop was uncharacteristic of the annual increase in Saudi holdings that had been monitored since 1975. International Monetary Fund statistics recorded a similar development; they showed a $10-billion drop in Saudi monetary reserves (to $19.7 billion) during this same period. According to the *New York Times,* "financial specialists identified some shift [in Saudi assets] into Canadian Treasury bills."[24]

Yet the cause of the Saudi shift was not explained by Saudi officials. Consequently, Washington was ripe with speculation that the Saudi move was politically motivated and not necessarily induced by economic considerations alone (such as diversification out of the dollar). It is inconceivable that such a political speculation could develop if a West European government, such as West Germany, began to draw down its reserves.

Despite a Saudi draw-down of its reserves in 1983 and 1984—needed to compensate for lower oil revenues—the political leverage available to Arab investors is still immense.

In early 1984, worldwide Arab assets were estimated to be in excess of $360 billion. The political leverage would disappear if most of the Arab financial assets were liquidated, but it would require a continuous oil glut for many more years.

On January 7, 1983, legislation was introduced in Congress to compel the executive branch to overhaul and improve its monitoring of foreign investment in the United States, especially with regard to acquisitions of "sensitive national interest sectors." Written by Rosenthal, the bill was an attempt to ensure that the Congress and American public be adequately apprised of the extent of OPEC investment in the United States. The bill, which never attracted any substantive support in the House of Representatives, was not introduced by Rosenthal, but by his friends in Congress: the day before, Rosenthal died after a long battle with cancer. And since that time, no other congressman or senator has been willing to monitor the issue of Arab investment.

17

THE SUPER LOBBYIST

It is the Israeli lobby's ability to politically reward or intimidate American
politicians and the media that has led
to such intractable support for Israel
among U.S. foreign policymakers.

*An excerpt from a
secret public relations
document written by
Robert Gray*

In Washington, Robert Gray is an important person. A very important person. He is the consummate Washington establishment figure par excellence, possessing that rare combination of access to the powers that be, unsurpassed public relations talent, and extraordinary popularity and prominence in Washington's social and political circles.

A former secretary to the Eisenhower cabinet, the Nebraska native now operates the most influential public relations and lobbying firm in the nation's capital. Gray started the firm in March 1981 after having spent twenty years as the chief Washington lobbyist for Hill & Knowlton, one of the world's biggest advertising and public relations firms, and Gray's company is now on the way toward becoming the largest independent (i.e., not owned by an advertising agency) public relations firm in the country.

With a staff of 130 lobbyists, publicists, and communicators, no other lobbying firm anyplace else has better, closer, or more extensive connections on both sides of the political aisle. Gray and Company's main headquarters is located near the beginning of the Chesapeake and Ohio Canal in Georgetown in a renovated electrical generating station immodestly called the Power House.

Gray himself has unparalleled connections to the Reagan administration and to the Republican party. A loyal friend, devotee, and supporter of Ronald Reagan, Gray was the director of communications for the Reagan-Bush campaign and wrote several widely circulated memos during the 1980 campaign detailing various strategies to defeat President Carter. After the election, Gray was made co-chairman of the 1980 Inaugural Committee and hosted a lavish black-tie dinner in honor of Nevada Senator Paul Laxalt, President Reagan's close friend and confidant. The sixty-two-year-old, elegant-looking, silver-haired Gray is on a first-name basis with every member of the Reagan cabinet, top White House officials such as Edwin Meese and Chief of Staff Jim Baker, and scores of senators and congressmen. With one phone call Gray can often schedule appointments on behalf of his clients with the chairmen of powerful Senate committees and the heads of executive branch agencies.

In order to broaden his access beyond the Republican party, Gray has hired a stellar cast of Washington politicos. These include Frank Mankiewicz, a veteran Democratic party aide who was press spokesman to Robert Kennedy and later a campaign adviser to George McGovern in his 1972 campaign (Mankiewicz later became head of National Public Radio, but resigned in 1983 amid a burgeoning financial mismanagement crisis); Alejandro Orfila, secretary-general of the Organization of American States; Gary Hymel, a well-known figure in Democratic party circles, who served eight years as chief of staff to Democratic Speaker Tip O'Neill; Bette B. Anderson, undersecretary of the treasury under President Carter; Joan R. Braden, a well-known Washington party-giver and former consumer affairs adviser for the Department of State; and Ronna Freiberg, former deputy assistant to President Carter for congressional liaison.

The credentials of the company's directors are even more impressive. They constitute a venerable Who's Who of the power elite in the United States. Among those who now serve, or have served in the past, are such illustrious names as Clare Boothe Luce, the first woman ambassador to Italy (1953–57), a former congresswoman, managing editor of *Vanity Fair,* prominent author, and legendary Republican party figure who now serves on President Reagan's Foreign Intelligence Advisory Board; Richard Wirthlin, Ronald Reagan's pollster; Bryce N. Harlow, former counselor to President Nixon; Melvin R. Laird, secretary of defense in the Nixon administration; and William Miller, secretary of the treasury under President Carter. In short, the firm is the very embodiment of Washington's post-Kennedy-Johnson political establishment: WASPy, rich, and bound together by the pragmatic cognizance that political bipartisanship is good for the pocketbook.

Gray offers his clients—who pay anywhere from $40 an hour for a secretary to $350 an hour for Bob Gray himself—a one-stop supermarket of services including lobbying on Capitol Hill, arranging tête-à-têtes with key senators and cabinet secretaries, setting up lavish receptions and parties around the country, contacting reporters and organizing media tours, devising extensive public relations and image-building programs, and even professional baby-sitting.

On behalf of Turkey, for example, which pays Gray an annual retainer of $300,000, Gray and Company lobbyists met with over four dozen senators, congressmen, and staffers on Capitol Hill over a six-month period in 1983. In the aftermath of Turkey's unilateral recognition of Northern Cyprus as an independent state, Congress had been very piqued. Turkey employed Gray to try to deter Congress from cutting military aid to Ankara. (Gray's efforts failed, however.) Another client is Martin Baker Aircraft Company of Britain, Ltd., which manufactures ejection seats for military aircraft. Gray and Company arranged meetings with members and staffs of six congressional committees with jurisdiction over defense matters, to have them consider prodding the Department of Defense to award contracts to Martin Baker for the F-18. For the National Broadcasting Company, organized a lobbying effort—which turned out to be unsuccessful—to pressure congressmen to vote in support of the network's proposals to syndicate television reruns.

Other Gray clients include the Bendix Corporation, the Tobacco Institute, Western Union, Sears World Trade, the John F. Kennedy Center for the Performing Arts, Stroh Brewery, Estee Lauder, Canada, the International Brotherhood of Teamsters, Warner Communications, and more than a hundred other corporations, law firms, foreign governments, unions, and professional associations.

Gray, the charismatic figure who carries himself with a suaveness that veteran Washingtonians still marvel at and whose popularity results in his wearing out two tuxedos every year, likes to be thought of as a bit more principled than the average public relations hired gun. His prospectus states that he "would not accept clients whose interests would be inconsistent with the national interests of the United States." In that regard, Gray and Company officials have told *Newsweek* and the *Washington Post* that they rejected Libya as a client. And, when asked by *Time* how he knew that increased military aid for Turkey was in the country's national interest, Gray responded, "I always check these situations with Bill Casey"—the head of the CIA, who is a personal friend of Gray.[1]

Gray also does pro bono work. He provided free publicity, for example, for a foreign policy conference featuring Henry Kissinger at the Georgetown University Center for Strategic and International Studies, and he has become involved in other prominent charitable efforts.

One of the first widely publicized events in which Gray demonstrated his civic goodwill occurred in the spring of 1982. In April, Wolf Trap, a popular music and performing arts center located in Virginia one-half hour from Washington, was destroyed by a devastating fire. The opening of Wolf Trap's annual summer season was threatened. A worldwide search was immediately conducted for a suitable temporary theater. Wolf Trap officials examined all possibilities, including a geodesic dome and an inflatable hockey rink. But none of these structures was appropriate. Finally, the existence of a unique modular tent was brought to the attention of Wolf Trap officials. Yet it was situated in a rather inaccessible location—the United Arab Emirates, where it had been used for a trade fair.

Suddenly, in mid-May, Gray—who at the time served as the head of Wolf Trap's board of directors—announced that Saudi Arabia had agreed to pay the $100,000 air shipping costs to transport the huge structure to the United States. Gray acted as the intermediary between Wolf Trap officials and the Saudi ambassador, conveying messages and helping to arrange the logistics. "Many countries here said they want to do something," commented Gray in an interview with the *Washington Post,* "but this is the first to step up and put its check on the line." The *Washington Post* added that Gray said that he had "no business relations with any Saudis, but is friendly with the Alhegelans [the Saudi ambassador to the United States and his wife]."[2]

Before long, a business relationship had commenced between Gray and the Saudis and other Arab interests. Moreover, Gray's work—though possibly routine for him—revolved around changing the course of American foreign policy and manipulating public opinion.

Even prior to the Wolf Trap event, Gray had begun working for Arab interests. In October 1981, the Kuwait Petroleum Company enlisted Gray and Company amid mounting congressional concern over Kuwait's purchase of Santa Fe International, a $2.5-billion oil and drilling company that also owned a highly sensitive nuclear technology subsidiary. For $10,000 Gray and Company counseled the client on "how to handle questions from the press," directed "American press inquiries" to their client, accompanied the head of the Kuwait Petroleum Company to a meeting with Congressman Benjamin Rosenthal, and "advised" the Kuwaiti official on his testimony before Rosenthal's subcommittee. Kuwait has continued to retain Gray's services for $60,000 a year.

In June 1982—one month after Gray declared that his relationship with the Alhegelans was one of friendship only—Gray signed a lucrative contract with an organization headed up by Nouha Alhegelan, the wife of the Saudi ambassador. In the wake of the Israeli invasion of Lebanon, Mrs. Alhegelan and the wives of other Arab ambassadors organized a group called the Arab Women's Council. The purpose of the new organization was to protest the Israeli actions and to relay to the American public the Arab side of the Israeli invasion.

For a fee of $300,000 for work conducted between June 15 and July 15, 1982, Gray agreed to arrange a vast public relations and political advertising campaign. This included designing, placing, and paying for full-page advertisements in national newspapers condemning the Israeli invasion and organizing an intense three-week media blitz across the country for Mrs. Alhegelan and other wives of ambassadors. Gray succeeded in setting up interviews with over sixty-nine radio and television stations and thirty-three newspapers, magazines, and wire services. The combined television audiences that heard Mrs. Alhegelan's pleas to "stop the genocide" in Lebanon totaled 19,500,000.[3] The full-page advertisements carried the headline "Begin's Holocaust in Lebanon." (The Arab Women's Council, which is funded by Arab governments, later became involved in a bit of chicanery when it secretly funded the money for a group called Peace Corps for Middle East Understanding. That group sent letters to 80,000 Peace Corps veterans asking them to lobby their congressmen to cut aid to Israel. To those who received the letter the group appeared to be making a sincere and independent effort to protest Israeli action. In fact, however, the letters had been mailed by the Saudi embassy, as evidenced by the numbers on the postage meter stamp.)

The business opportunities presented by the Israeli invasion of Lebanon, beyond Gray's contract with the Arab Women's Council, were apparently too good to pass up. According to informed sources, Gray submitted a twenty-nine-page confidential proposal to the National Association of Arab Americans, one of the principal pro-Arab lobbying groups in the United States.

In the document, Gray proposed "a national telethon for the victims of the Israeli holocaust in Lebanon, targeted at 25 major markets and scheduled over the course of a year [that] is the most ambitious and far-reaching proposal to date designed to alter American attitudes on the Middle East."

In the secret proposal, a copy of which I have obtained, Gray impressed on his prospective clients the urgency of buying his proposal:

> *The Lebanese crisis and the subsequent erosion of public support for Israel, does not represent so much a victory as an opportunity. . . . Now is the time to capitalize on the opportunity that has been created. . . . Gray and Company will provide, upon request, the National Association of Arab Americans with public relations/public affairs services designed to advance NAAA's objectives, public visibility, and impact upon U.S. attitudes and policies.*

Much of the remainder of the Gray document explained why a "nationwide television/telethon campaign" was the most effective way of changing American attitudes on the Middle East. Included was a discussion of previous nationwide telethons—including a mention of the success of Jerry Lewis's annual muscular dystrophy telethon—and a detailed presentation of production, television, and related costs. Though the telethon was to be organized around "the relief of Lebanese war victims," Gray makes clear that the underlying objective was to weaken "public support in this country for Israel." Gray even offered in his proposal to help form a "broad umbrella group of organizations"—in effect, a front group—which would sponsor the telethon to make sure it was "well received by U.S. television broadcasters."

Gray impressed on the National Association of Arab Americans the urgency of "beginning this project as soon as possible while memories of events in Lebanon are still fresh in

American minds.'' He promised that the $2-million telethon would be ''sufficiently hardhitting so as to have an important impact on the American public long after the fighting in Lebanon had ceased.''

Though the National Association of Arab Americans did not follow up on the telethon proposal, the group did retain Gray for other services. According to Gray and Company officials and other sources in Washington, Gray's work involved advising on a national advertising campaign to cut foreign aid to Israel. Gray and Company helped devise newspaper, radio, and billboard advertisements and placed them in cities as far apart as Topeka, Kansas; Albany, New York; San Mateo, California; and Little Rock, Arkansas. On one NAAA billboard in Topeka, for example, the words ''Lebanon '82'' were shown dripping in blood with a caption underneath: ''Is this how we want our tax dollars spent by Israel?''

Throughout the remainder of the year, Gray expanded his work on behalf of the Arab lobbying group, helping to design the copy for other radio advertisements across the country. Within several months' time, Gray's work on behalf of this organization was no longer a secret in Washington.

Suddenly, however, in December 1982, Gray and Company announced that it was terminating its relationship with the lobbying organization. Why? Because, said one company official in an interview with the *New York Times,* the National Association of Arab Americans was ''too strident.''[4] ''In short,'' reported *The Times,* ''[NAAA] rejected most of their advice and insisted on an anti-Israeli campaign that could not be reconciled with the company's policy of not doing anything in foreign affairs that goes against official American policy.'' In other interviews, officials claimed specifically that they had been against advertisements in Pennsylvania linking unemployment to aid to Israel. Moreover, in a bizarre twist, Robert Gray discreetly drove to the Israeli embassy around Christmastime 1982 where he met with Israeli Ambassador Moshe Arens. According to sources privy to the meeting between the two men, Gray told the Israeli ambassador that he harbored no hostile feelings toward Israel and that he was forthwith desisting from any anti-Israeli public relations campaign.

In fact, however, a mammoth public relations proposal authored by Gray was still being circulated with Saudi Arabia and American-Arab organizations that, in the opinion of one highly placed official in the Arab lobby, contained "raw Jew-baiting." Entitled "A Strategy to Improve Perceptions of Arabs in the United States," the secret eighty-one-page proposal contained a massive "communications" laundry list. This included a coast-to-coast media campaign, press liaisons, a speakers' program, nonprofit front organizations, symposia, television and newspaper advertising, posters, bumper stickers, and "everything else," said this source, "that had been proposed by other Arab lobbyists for the past five years." The proposal contained more than a dozen separate sections that could be implemented independently and was to cost up to $9 million for two years of operations.

In the document, excerpts of which I obtained in January 1983, Gray portrayed support for Israel as exclusively the result of the "Israeli lobby's" manipulations: "It is the Israeli lobby's ability to politically reward or intimidate American politicians and the media that has led to such intractable support for Israel among U.S. foreign policy makers." There was no mention at all of the possibility that support for Israel might also be the product of genuine popular sympathy for Israel. The end result of Gray's proposal served to reinforce the notion that a conspiring "Zionist" cabal controls American public opinion, the media and Congress.

The document continued:

The Israeli lobby's influence, however, goes well beyond its ability to "buy" friends in Congress with votes or political contributions. There are many other groups in America with larger and more widely dispersed constituencies which even outspend pro-Israeli organizations. Why, then, is the Israeli lobby so effective? Because those who control the Israeli lobby know how to wield power and use it effectively.

The techniques of manipulation of American Middle East policy have been polished and expanded over the last 40 years. Tens of millions of dollars annually are spent in public education and political activities by the American Israel Public Affairs Committee, B'nai B'rith, the American Jewish Congress, Hadassah, and other pro-Israeli groups.

Were it not for one other important tactic, even all this would be inconsequential: the Israeli lobby never lobbies for Israel. All of the organizations in the United States which comprise the Israeli lobby, "wrap their arguments in an American flag." They relentlessly present Israel's national interest. Their success in connecting Israel's national interest to U.S. national interest hides the inherent conflict between the two nations and, until now, has prevented a fair debate of the issues.

The document describes how AIPAC (the primary pro-Israel lobbying organization) has operated "a sophisticated classification system." Gray also wrote "that even in districts where there is no significant Jewish voting bloc, candidates are often afraid of being outspoken against Israel for fear of retribution from the Israeli lobby which does not hesitate to fund an opponent."

According to sources connected with Gray's firm, the document was prepared sometime in the fall of 1982, soon after Gray had landed a $100,000 a month contract with one of the most senior members of the Saudi royal family. He is Prince Talal Abdul Aziz ibn Saud, the twenty-third son of the founder of Saudi

Arabia, half-brother of King Sahd and third in succession to the throne. Talal retained Gray for nonpolitical work—to assist the prince in his capacity as special envoy to UNICEF. These activities included organizing the prince's cross-country tours on behalf of UNICEF, providing media relations, setting up fund-raising receptions and dinners, and even producing favorable films about the prince.[5]

After Talal retained Gray on August 1, 1982, Gray prepared "a strategy paper" at Talal's suggestion on how to sway American public opinion away from Israel. The paper was to be presented to Talal's brothers, the leaders of Saudi Arabia. Gray prepared a massive proposal and submitted it to Talal who then forwarded it to the royal family. In addition, Gray even traveled to Saudi Arabia at Talal's invitation, to meet with several members of the royal family to discuss his proposal.

Gray told one organization in Washington that he spent $30,000 of his own money to prepare the plan. In the end, the royal family did not hire Gray to execute the proposal. Yet various components of the secret document may have been implemented in 1983, according to a senior official in the Arab lobby.

If this is so, it is not known who footed the bill. One possibility, however, and this is speculation, is the Arab Information Center, an arm of the League of Arab States. In May 1983, Gray registered as a foreign agent for this organization, which is funded by Arab governments. According to his registration, Gray's work has involved analyzing "Arab presentations in the U.S. media," consulting on public relations, and a major research paper involving Latin America. In the next six months, the League of Arab States paid $110,000 for Gray's services. Over $70,000 of this money went to pay for unidentified "consultants" working on unspecified projects on behalf of Gray.[6] A source privy to Gray's work on this account said that Gray had undertaken to analyze the media, particularly newspapers, "hostile to the Arab cause" and to implement or suggest "corrective public relations measures." One of the newspapers which has been intensely scrutinized and evaluated for the Arab League is the *Wall Street Journal,* whose editorials have been deemed by the Arab League to be "anti-Arab."

In 1984, Gray terminated his relationship with the Arab League and ceased all public relations activities relating to the Arab-Israeli dispute.

18

THE WEST CONNECTION

I don't like to be accused of being
bought with Saudi money.

John C. West

In his folksy southern drawl, the guest speaker at the eight o'clock breakfast meeting captured everyone's attention. "Our credibility in the Arab world is the lowest it's ever been," he informed the congregants. He then fiercely denounced Israel, and relayed the charge by Saudi leaders that Israel had wantonly engaged in the "massacre of men, women, and children in Lebanon." On that spring morning, April 14, 1983, on Hilton Head Island, a posh forty-two-square-mile resort off the coast of South Carolina, the members of the First Presbyterian Church had gathered to hear a local fellow who had made good. John Carl West, their former governor who had risen to international prominence as American ambassador to Saudi Arabia, came to speak about the "troubles of the Middle East."

Moreover, the speaker wanted to impart his special knowledge of the vast control exerted by the Jewish lobby across the United States—a fact, he said, his Saudi Arabian friends were well aware of: "It is a ritual that any presidential aspirant must pay homage to the concept of Israel and its objectives, or they will not be supported by this powerful lobby. As many elections for Congress and the Presidency are decided by slim margins, this can be crucial.'' He charged that the presidential candidacy of John B. Connally was automatically doomed because of Connally's Middle East peace proposal, adding: ''If President Reagan wants to be reelected, I would advise him not to say anything against Israel until after the 1984 election.''

During his ninety-minute talk, he painted Israel as an out-of-control rogue state bent on sowing the seeds of regional conflict. The Soviet Union had been able to establish influence in the Middle East, West said, primarily because of Israel and its alliance with the United States. Just before the meeting disbanded, a member of the audience raised a question about an ''Israeli massacre'' of ten thousand Christian Lebanese. In response, West told the crowd that, yes, he understood that Israel had participated in such a massacre but intimated that the press had suppressed the news of it so that it never reached the American public.[1]

The fact of the matter was that this massacre never took place. No one has heard of it. West himself, in an interview with me over a year later, denied any knowledge of any Israeli ''massacre or atrocity.'' But his allegation before the Presbyterian church group fit perfectly with the rest of his speech. Indeed, his portrait of a murderous Israeli government working with an omnipotent Jewish lobby had become standard fare in his public speaking. Ten months before, he compared the Israeli offensive against the PLO in Lebanon to Hitler's liquidation of millions of Jews. ''Hitler tried that with the Jewish situation,'' he told a newspaper interviewer in an analogy whose corollary suggested, somewhat cryptically, that the Jews in World War II Europe had created a problem.[2]

Though it is impossible to gauge his precise impact, West's views were surely given popular credence, especially since his credentials after a lifetime of government service are impeccable. West's speech before the Presbyterian group, for example, produced two lengthy articles in the two Hilton Head newspapers, quoting his comments extensively and uncritically. Years ago, West had made a name in the newspapers for different reasons: he was known for his passionate commitment to social, economic, and racial justice, reminding some South Carolinians of John Fitzgerald Kennedy.

Born on August 27, 1922, West was raised in Camden, South Carolina, a city of ten thousand, located an hour and a half east of the state's capital, Columbia. He was reared by his mother—his father died in a tragic fire when West was a small child—on a 220-acre farm. He received a scholarship from the Citadel, the national military college in Charleston, South Carolina, where he graduated in 1942. After serving four years in Army military intelligence, West returned to school, obtaining his law degree from the University of South Carolina Law School in 1948, graduating magna cum laude.

Six years later, West began his foray into state politics; he was elected to the South Carolina State Senate by a margin of just three votes—out of eight thousand. As a state senator, he represented rural Kershaw County from 1954 to 1966, during which time he championed the cause of education. An extremely amiable man, West never forgot his humble background. He was known to be ambitious, but not ruthless. West's rise to power was enhanced by his shrewd political savvy, which allowed him to defeat his opponents without destroying them.

Except for the four years he spent in South Carolina's governor's mansion between 1971 and 1975, West lived in his hometown of Camden, where he was a country trial lawyer in the law firm he had started in 1948—West, Holland, Furmand, and Cooper. A charismatic, able politician, West, said one veteran state senator, "could always tell an old Abe Lincoln story." His background, however, seemed hardly appropriate for one of the most important and sensitive diplomatic posts in the world— the ambassadorship to Saudi Arabia.

Back in South Carolina people were proud but bemused when the news reports reached them in February 1977 that West was likely to get the appointment. They knew, of course, that West was in line for a high-level federal position. As a nationally respected governor, West had provided his good friend Jimmy

Carter with an invaluable boost by declaring his early support for the dark-horse presidential candidate in early 1975. Even before Carter won the election, West had been one of the privileged few to work confidentially on Carter's transition team led by Atlanta lawyer Jack Watson. The team had busily been making tentative staff recommendations since June 1976, five months before the election.

Just one day after Carter was elected, the authoritative *Kiplinger Washington Letter* revealed that West was being considered for the secretary of commerce slot in the new cabinet.[3] Two South Carolinians had occupied that post during the previous forty-five years: Daniel C. Roper from 1933 to 1939 under President Roosevelt, and Frederick B. Dent from 1973 to 1974 in the Nixon-Ford administration. But West was not appointed, later telling friends and acquaintances that he had not been interested in the position and was not asked to serve in it. Some of West's political cronies in the South Carolina State Senate thought that West would be appointed to a federal judgeship, but that did not materialize either.

As the Carter administration assumed office in January 1977, it appeared that West had still not collected his IOU. Some South Carolina legislators at that point thought that West was simply going to follow in the footsteps of his predecessor, former Governor Bob McNair who, upon his retirement, established a very lucrative legal practice.

But, unbeknownst to South Carolinians, immediately after Carter was elected, West had met with a former secretary of state, Dean Rusk, newly designated Secretary of State Cyrus Vance, and other soon-to-be Carter officials at a political forum at the Citadel. "John," said Rusk, "the President needs people who have been his friends and associates."

"I really don't see any job in Washington that I want," West responded, adding, "There is only one job that I would want."

"What's that?" asked Rusk.

"Ambassador to Saudi Arabia," said West to a surprised Rusk.

His request was relayed to the President and granted. On the morning of Tuesday, May 24, the White House officially announced the appointment. That same day the new President welcomed Crown Prince Fahd and Sheikh Yamani to a series of meetings.

Reaction to West's appointment was positive. South Carolina senators Democrat Ernest (Fritz) Hollings and Republican Strom Thurmond announced their enthusiastic backing. Yet the fact that West was the first noncareer foreign service diplomat ever to fill the post of ambassador to Saudi Arabia evoked surprise on Capitol Hill.[4] Why would a southern country trial lawyer turned governor, his political experience limited to state affairs, be interested in such a position?

In reality, West's yearning for the position was not surprising, given the new direction his life had taken as a result of the petrodollar boom.

Elected governor in 1970, West had been hailed as one of the new breed of southern progressives, along with his Georgian counterpart, Jimmy Carter. West promised to initiate immediately a series of programs to end hunger, expand educational opportunities, and improve housing. Before a crowd of six thousand at the state capital, West gave a stirring address pledging to eradicate "any vestige of discrimination" and to "break free and break loose of the vicious cycle of ignorance, illiteracy, and poverty."[5] Two of his first acts as governor were to appoint the first black staff member to the governor's office and to create the state's first Human Affairs Commission to combat racial discrimination. In the fourth year of his term, West vetoed the popular death penalty bill, explaining in very moving terms that the death penalty was the ultimate in "physical torture." By the end of his four years, however, dissatisfaction with West arose from those who felt he was not progressive enough and from a larger segment of the population who felt he had gone too far.

But there was unanimity about one area of West's accomplishments. In the mid-1960s South Carolina had embarked on an aggressive campaign to attract investment in local industry from out of state and foreign sources. Lured by tax incentives and low labor costs—there are relatively few unions in South Carolina—companies and investors from New York and as far away as Germany, France, and Great Britain began to relocate or establish subsidiaries in South Carolina. The South Carolina State Development Board sponsored numerous trips to Europe and the Far East.

The efforts paid off magnificently for South Carolina. West German investment in the Palmetto State may be the highest in the United States. On a per-capita basis, foreign direct investment in South Carolina is the second highest, after Louisiana, of all the fifty states. Sleepy backwater towns were transformed into dynamic urban centers such as the area surrounding and including Charleston, a vibrant harbor city on the Atlantic coast.

Though the state already had a great number of textile mills, the influx of out-of-state investment led to the emergence of South Carolina as the chief textile-producing state in the country. Most of the mills, factories, and synthetic fiber plants were concentrated in a relatively small area in the northwest portion of the state known as the Piedmont Plateau, clustered in and around the cities of Anderson, Spartanburg, and Greenville and along Interstate 85. So much West German money poured into this area that Interstate 85 has been jokingly referred to as "the Autobahn." Industrial development sprouted throughout the state.

The 1973 Arab oil embargo hit the economy of this area particularly hard. Acutely linked to the natural decline in discretionary purchases by the average consumer across the United States, South Carolina's textile and apparel industry, which provided over 40 percent of the state's employment, was hurt by massive layoffs creating a recessionary multiplier effect throughout the state.

However, as the Arabs started to amass fantastic wealth, they looked for new places to invest their money. An American financial adviser to Kuwait, who had previously lived in South Carolina, suggested an undeveloped tract of land fifteen miles off the coastal city of Charleston. So, in 1974 Kuwait purchased Kiawah Island, a beautiful, isolated, and idyllic ten-thousand-acre retreat. The price tag was $17.5 million, and it represented one of the first major real estate investments in the United States by an Arab country. Two years later it opened as a major tourist resort. Its current value is considered in excess of $100 million.

The Kuwaiti investment could not have come at a better time for South Carolina. West, whose tenure as governor saw the influx of an estimated $1 billion in out-of-state investment, soon became a believer in the miracles afforded by the unlimited riches of the Arab world. Having befriended Kuwaiti officials in South Carolina, West decided to accept their invitation to visit the Middle East.

Right after Christmas 1974, barely three weeks before the expiration of his term, West and his wife traveled to Kuwait at the expense of the Kuwaiti Investment Company. He planned to hand his Kuwaiti hosts a veritable Christmas shopping list, including suggestions for Kuwaiti investment in South Carolina oil refineries, Kuwaiti purchase of textile products, and Kuwaiti utilization of pollution control and agribusiness expertise offered by South Carolina's colleges.

During the trip, a particular political disposition critical of American policy in the Middle East began to emerge in West— just one year after the oil embargo and price hike had so battered his state. In his discussions with the Kuwait foreign minister, West was asked about Kissinger's threat to employ military force to ensure a secure supply of oil "when there's some actual strangulation of the industrialized world."

Upon his return, West relayed to reporters the substance of the comments he had made in Kuwait: "I told him [the Kuwaiti foreign minister] . . . I did not think our country would tolerate an act of aggression against another country because of selfish economic considerations."[6] And he told his Arab hosts that it would be a mistake to equate "a few northern journals, such as the *New York Times* and the *Washington Post"* with "the ground-root sentiment of the average American and average South Carolinian." The South Carolina governor also told reporters that the "first question [he] was asked by one foreign minister was 'Can you tell me what any Arab has done to Americans to cause them to hate Arabs?' "[7]

West made these comments in his last press conference as governor in early January 1975, in which he waxed ecstatic about the "great" opportunities for Arab investment in South Carolina and strongly urged the state to aggressively seek petrodollars. "Governor John C. West sees the Mid East as the new financial capital of the world, and he sees no reason why South Carolina should not try to attract some of that money," reported the *Columbia Record* the day after the press conference.[8] West's exhortations were soon heeded by his successor, Governor James B. Edwards, who later became secretary of energy in the Reagan administration.

In August 1975 a delegation of Arab officials visited South Carolina. Kuwait's foreign minister of finance and ambassador flew to Columbia where they were feted at a reception hosted by Edwards and West. The head of Saudi Arabia's postal service and high-ranking Saudi educational officials also traveled to South Carolina to discuss plans for training Saudi postal employees and to increase Saudi enrollment in American universities. Three days of meetings took place with officials of the University of South Carolina, the State Development Board, the U.S. Postal Service, and local companies. It was an instantaneous love affair. At a luncheon in honor of the visitors hosted by former governor West, it was disclosed that a South Carolina trade mission would visit the Middle East within the next year to look for additional contracts.[9] The idea for the mission had come from West.

Sponsored by the State Development Board, the fifteen-man trade mission, headed by West, departed for Jordan, Kuwait, and Saudi Arabia in April 1976. A small furor, however, erupted several days before the group's departure when the Jewish community of Charleston protested the mission's apparent compliance with Saudi Arabia's exclusion of Jews. Members of the mission were required to provide proof of their non-Jewish religious background by submitting a letter from their church or a baptismal, birth, or marriage certificate revealing a church affiliation. The acting director of the State Development Board, F. Earl Ellis, defended the trip: "We feel that the purpose of the mission is very legitimate from the standpoint of generation of development in South Carolina." State Port Director W. Don Welch dropped out of the mission because he was offended by the visa policy, but the rest of the members departed as planned.[10] West, though out of office, was received by the Saudi royal family as well as by King Hussein—evidence of his growing favorable reputation in Arab capitals.

"The new frontier of the world is the Middle East," West expansively observed in a speech to southern businessmen on his return.[11] The trade mission bore fruit for West beyond the anticipated $14 million in Middle East contracts that were firmed up for South Carolina businesses. On May 16, 1976, just twelve days after he returned from the official trade mission, the former governor announced that he was forming a new company, a sort of one-stop consulting service for American firms trying to land petrodollar contracts with Arab nations. Called Arab-American Development Services, Inc., West's new company specialized in providing "entrée for business firms in the Middle East," such as handling negotiations with Arab customers, scheduling appointments, and even providing interpreters.[12] The other partners in the firm were Sabah-al-Hay, editor of the *Arab Economist;* Chafic Akhras, an international banker; and Joseph J. Malone, a self-described investment banker and financial consultant who specialized in the Middle East. West also revealed that his son, Jack, would depart for the Middle East to help solidify potential Arab clients.

On June 16, 1977, West was sworn in as ambassador to Saudi Arabia. But within five days, a political transformation on his part had already become evident. He declined to be present for the White House ceremonies on June 22 where Carter signed the new anti-boycott legislation, and he even refused Carter's offer of the pen he had used.[13]

Yet three weeks before, in an interview with *Columbia Record* reporter Jan Stucker, West had a decidedly different perspective: he professed his unconditional support for congressional legislation designed to counter Arab blacklisting of American firms that employed Jews or firms that traded with Israel under the Arab boycott. Any boycott, he said, based on "race, religion, color, or national origins is morally wrong and ought to be declared morally wrong. And any legislation that would accomplish that I would support wholeheartedly."[14]

Within months, the agenda of the American embassy in Saudi Arabia reflected West's heavy pro-business orientation. Hordes of entrepreneurs, businessmen, and consultants—a disproportionately high number of them from South Carolina—descended on the kingdom. "We ought to initiate a direct local flight from Columbia to Jidda," commented one local businessman. Some came to solidify contracts; many others arrived to seek new ones, hoping to take advantage of the doors opened by West.

A local Democratic party activist, W. W. (Hooty) Johnson, head of Bankers Trust, traveled to the kingdom without any "specific mission," he acknowledged in the *Columbia Record,* adding, "When we have an ambassador in what is called the richest country in the world, naturally you want to explore any opportunities that might be there."[15] Donald Fowler, chairman of the South Carolina Democratic party, who also operated a public relations firm, accompanied a bus manufacturer from Arkansas who was looking for additional business with Aramco. Flexi-Wall Systems, a Greenville company, landed hundreds of thousands of dollars' worth of contracts in Saudi Arabia and other Persian Gulf nations for its patented wall-surfacing materials. Even West's son became a frequent business visitor, according to an official in the American embassy.

One contract, however, provoked a major controversy, exploding on the front pages of South Carolina newspapers and the *New York Times*. On March 27, 1978, the public relations and political consulting firm of Cook, Ruef, Spann, and Weiser, was awarded two Saudi contracts worth $165,000 to lobby for the sale of F-15s and for a comprehensive public relations campaign. J. Crawford Cook, the forty-five-year-old head of the firm, received the contract on his third visit in six months to the kingdom. West provided Cook with invaluable entrée to the Saudis. The two men were very close politically as well as socially. In the mid-1950s Cook, a UPI reporter, had gotten to know State Senator West. Later, Cook became West's campaign manager in 1970, and Cook's firm retained West as its attorney following the expiration of the governor's term.

Cook beat out several leading New York public relations firms for the contract with the Saudi government. He acknowledged the help of his friend. "There is no way to minimize the importance of John telling the Saudis, 'Yes, he's a friend of mine.' "[16] But Cook denied any impropriety on West's part.

Yet, in Washington, the Cook-West connection raised questions. Moreover, congressional suspicions of West were heightened by actions that some thought displayed questionable professional judgment. The ambassador engaged in especially strenuous lobbying on Capitol Hill on behalf of the sale of the F-15s to Saudi Arabia. Moreover, in one of his briefings to the Senate Foreign Relations Committee staff, West went well beyond administration statements, bluntly warning that the price of Saudi oil would increase if the sale of the F-15s did not go through.

Investigators for Congressman Benjamin Rosenthal's Subcommittee on Commerce, Consumer, and Monetary Affairs also discovered that West had signed an American embassy report containing a "flattering" profile of Saudi businessman Adnan Khashoggi. Though American embassies routinely prepare such background reports on foreign sales agents for American businesses, the report on Khashoggi conspicuously omitted the fact that he had figured prominently as the middleman in the payoff scandals involving Northrop, Raytheon, and other multinational corporations, and that he had refused to comply with the Securities and Exchange Commission's subpoena.

As the details surrounding Cook's new contract began to filter back to Washington, an uproar ensued. *New York Times* columnist William Safire blasted the "Ambassador's business fixing [as] grossly unethical" and called for an investigation by the Senate Foreign Relations Committee.[17] Within days, Kansas Republican Senator Robert Dole reiterated the call for a Senate investigation: "It is alarming to think that an Ambassador would consent to perform a liaison function between commercial interests in this country and the government to which he has been assigned."[18] In South Carolina, the *Columbia Record* roared in a three-inch headline on April 3, "Ex-Governor Embroiled in Growing Firestorm."

Interviewed for that article, West adamantly denied any wrongdoing: "I haven't done anything I am ashamed of, and I would do it all over if I had to do it again. . . . Just because people are my friends, I don't think I should have to turn a cold shoulder."[19]

A State Department spokesman hurried to West's defense: "We have in this case no reason to believe there is any impropriety in the action of any U.S. government official."

West had also irked some people by his passionate defense of Saudi Arabia's "moderate" policies, especially since he offered that defense right after Saudi Arabia praised the PLO killing of thirty-one Israeli civilians in an attack on an Israeli bus. Saudi Arabia called the attack a "courageous operation." When a reporter for the Charleston-based *News and Courier* asked him about Saudi funding of the PLO, West said the Saudis "believe Arafat is moderate and they support him. . . . I would feel very, very confident that they would not knowingly support any group that sparks a terrorist raid." West also told the interviewer that he advised the Saudis that the "PLO does not have a good image.'' In that same interview, West indicated that his own credibility with the Saudis was at stake.[20]

Soon, both Cook and West proved their credibility with the Saudi royal family. On May 16, the Senate approved the sale of F-15 fighter aircraft to Saudi Arabia by a vote of 54 to 44, a margin of victory that surprised even the winning side. West, with South Carolina Senator Hollings at his side, had personally lobbied senators to support the weapons sale.

Within a very short time, West had developed an extremely close rapport with the Saudi royal family. And, according to a former senior official in the Carter White House, West ''would often convey personal messages from President Carter to the king and members of the royal family.'' But West's close relationship was accompanied by what some considered an almost blind embrace of Saudi policies. Nowhere was this more evident than in the critical period following the tentative approval of Egyptian President Anwar Sadat and Israeli Prime Minister Menachem Begin of the Camp David Accords negotiated by President Carter in September 1978 at the presidential retreat in Maryland.

Because the Saudis provided billions of dollars in aid annually to bolster the Egyptian economy and because Saudi subsidies also commanded substantial leverage over Jordan, the Carter White House deemed absolutely vital Saudi Arabia's support of the Camp David Accords. According to a senior State Department official, Sadat himself became anxious about the Saudi reaction during the marathon negotiating sessions at Camp David, but he was reassured that Washington would guarantee the forthcoming support of the Saudis. Secretary of State Vance was immediately dispatched to Riyadh following the provisional agreement by the two Middle East leaders in September.

But his efforts were to no avail. Saudi Arabia would not lend its support. The Carter White House was clearly upset by the lack of Saudi support for the Camp David Accords. ''In the U.S. perspective, Saudi Arabia is as much the molder of the Arab consensus as its captive,'' wrote *Washington Post* diplomatic correspondent Don Oberdorfer. "The Saudi failure to support or even to acquiesce in the Egyptian-Israeli treaty, in this view, saps its chances for success."[21]

Within the next two months, Saudi leaders—who six months before had received permission to buy sixty F-15s—chastised Sadat, condemned the Accords, and helped organize the Pan-Arab effort to pressure the Egyptian leader to annul the peace treaty. At a Pan-Arab summit held in December 1978 in Baghdad, $3.5 billion was offered to Sadat to renounce the Accords and additional billions were promised to Jordan and Syria if they would continue to boycott all peace negotiations. Following the treaty-signing ceremonies in March 1979, another Arab summit was held, this time resulting in more severe sanctions applied to Egypt, including termination of all foreign aid, a trade boycott, and the severance of diplomatic relations. Once again the Saudis provided the critical consensus.

In an interview with me, West maintained that he "explained to the Saudis the advantages" of supporting the Camp David Accords, but "they would ask me this question which was difficult [for me to answer]: 'You show us what advantage it would be to the peace process to have our support?' "[22]

The full role played by West as the United States attempted to elicit Saudi support for the Camp David Accords will not be known until all classified cables are made available. But, according to one high-ranking official at the time, the American ambassador to Egypt, Hermann Eilts, a career diplomat, and according to a foreign service officer who worked for West, West was far more disturbed by the Camp David Accords and Sadat's policies than he was by Saudi obstructionism.

West has adopted the "Saudi position hook, line, and sinker," said Eilts in his embassy office in Cairo in December 1978 to an astonished four-member Senate Foreign Relations staff delegation, which included me. (We were in Cairo to assess American economic assistance.) Eilts bitterly complained about West, asserting that West has operated a "public relations firm for the Saudis." His blunt assessment of a colleague's performance was virtually unheard of in the rigid world of foreign service etiquette. Yet, the substance of his criticism was repeated to me by a foreign service officer who worked under West for two years in the American embassy in Riyadh. West would routinely criticize the Camp David Accords and Sadat during staff meetings. Moreover, whenever Sadat lashed out at his Arab and Saudi critics—Sadat charged that the Saudis had pressured other Arab countries to break relations with him—West would be especially harsh in his denunciation of the Egyptian leader.

During the better part of one particular staff meeting on December 10, 1979, recalled the foreign service officer who kept his notes from that meeting, the Saudi ambassador launched into a "vituperative" personal attack on Sadat, who had just criticized the Saudis for their refusal to support the Camp David Accords: "Why can't Sadat keep his big mouth shut? Because Sadat has a personality problem." If this is what West is telling his own embassy, thought the foreign service officer, then "what is he saying to the Saudis?"

In newspaper interviews and speeches, West often bemoaned Western criticism of the Saudis or portrayed their oil production and pricing policies—even when the price of oil was rising precipitously—as sacrifices to the United States. For example, during an OPEC meeting in Abu Dhabi in December 1978, the Saudis joined in raising their oil prices by an economically jarring 14.5 percent. Seven months later, at the height of the gasoline shortage in the United States, the Saudis increased production by a million barrels but also raised oil prices by a staggering 61 percent, prompting the *Washington Post* to blast the Saudis and the Carter administration for portraying such actions as deriving from Saudi Arabia's "historic friendly relationship with the United States." "Nonsense!" exclaimed the *Post*. "The truly 'friendly' act would have been to pump a large quantity of oil, at a price closer to the cost of production. For the Saudis to join in raising the price of oil a crushing 61% in this year alone, and then to be thanked for agreeing to sell more makes the head spin." [23] By year's end, the Saudis had allowed the price of oil to spiral to $26 a barrel. Over $59 billion in oil revenues flowed into Saudi Arabia's treasury—the highest annual amount it had ever received.

For West, these actions represented Saudi altruism and friendship for the United States. "They [the Saudis] need production of half that amount [they are currently producing] to produce all the cash they need for their five-year development plan," he told the Forum Club in Houston on April 12, 1979, as he commended the Saudis for meeting Western needs and not charging as much as they could. "What we need to do and what is essential is to persuade Saudi Arabia to keep up excess production." West also lamented public criticism of Saudi Arabia: "The Arabs are very sensitive, the Saudis particularly, to newspaper reports." He added, "Every time they see a statement critical of their oil prices or policies, they take it to heart."[24]

A year later, when the American economy was still reeling from the shock of oil price hikes, West, on a visit to the State Senate in Columbia, South Carolina, continued to praise Saudi oil production policies. He claimed that the Saudis were producing twice as much as they really wanted. "They point out very vividly that a barrel of oil in the ground is worth more than any investment because of inflation," he said, pointedly ignoring that it was the rise in oil prices that had caused the inflation.[25]

When the *Death of a Princess* episode exploded in 1980, West quickly echoed the Saudi line. The same man who as governor had so courageously vetoed the proposed South Carolina death penalty as a "barbaric, savage concept of vengeance which should not be accepted, condoned, or permitted in a civilized society" now fiercely defended the Saudi regime that had shot a nineteen-year-old princess and beheaded her lover for committing adultery. When West was informed that South Carolina Education Television had canceled the show, he called the station to applaud its decision.

His total political conversion to the Saudi cause appeared complete by the time he left office, as evidenced by the news conference he gave to American reporters the month before he ended his nearly four-year tenure on March 1, 1981. In his final parting shot, West warned the newly elected Reagan administration that it had to shore up "the special U.S.–Saudi relationship" by making "progress that is discernible to the Arabs" on the Arab-Israeli conflict.

And in almost apocalyptic terms, West warned that "time has run out," that Saudi oil production would plummet, and that the American economy would suffer.

He conveniently added, contradicting what he had just said, that President Reagan could "buy instant credibility and time" if he agreed to sell to Saudi Arabia the bomb racks, fittings for the Sidewinder air-to-air missiles, and additional fuel tanks, all offensive equipment that had been denied to the Saudis in 1978.[26]

Once out of office, West began championing the Saudi political cause at public forums. When a commentator wrote in the *Wall Street Journal* that Saudi Arabian oil production policy was based on economic self-interest and did not depend on whether a pro- or anti-Western elite ran the government, West hotly contested the writer's thesis in a letter to the editor: "Saudi Arabia's concern for [the] free world economy and its friendship with the United States" was responsible for increased oil production.[27] When Israel attacked Iraq's nuclear reactor in July 1981, West sent a letter to the House Foreign Affairs Committee asking that punitive action be taken against Israel.

As a guest on NBC's "Tomorrow," West asserted that the key to Middle East peace was in negotiating with the PLO.[28] And during the eight-month debate over the sale of AWACs, West, according to Senate and South Carolina sources, lobbied intensively for the sale of the advanced radar planes, testified before Congress, contacted governors, senators, and American businessmen, and provided advice to Saudi officials and their registered agents in Washington.

In his numerous speeches across the country to business groups, lobbying organizations, and religious gatherings, West routinely speaks about the influence and power of the American Jewish community. His audiences are often told that Alexander

Haig approved the Israeli invasion of Lebanon and that the former secretary of state received a $20,000 fee from B'nai B'rith for a speech he delivered after he left office, implying that Haig was bought with Jewish money. Also, West never fails to attribute the decline of John Connally's presidential bid to Connally's Middle East "peace" blueprint, despite the fact that less than 10 percent of American Jews are registered Republicans, that they are not big contributors to the Republican party, and that Connally raised $10 million for his campaign anyway.

In an address before the First Presbyterian Church in Hilton Head in April 1984, one year after his first speech to the same group, West extended his portrait of the all-powerful Jewish lobby.[29] He recounted to his 150 listeners—many of whom were retired business executives and generals, and some of whom were former CIA officials—how Presidents Carter and Eisenhower were purportedly warned by their advisers that if they pressured the Israelis, they would "alienate the Jewish vote in New York and lose the presidency." Asserting that the population of American Jews is "fourteen million"—actually, six million is the correct number—West predicted they would all vote as a "bloc" on the Jerusalem issue—the legislation before Congress that would support moving the American embassy from Tel Aviv to Jerusalem, Israel's capital.

The extent of West's support for the Saudis has exceeded political advocacy. In April 1982 Saudi Prince Turki bin Abdul Aziz, sixth in line to the throne, became embroiled in a violent altercation with Miami police. Until 1979 Prince Turki had served as deputy defense minister of the Saudi government, but he then "resigned" after the royal family reportedly looked with disfavor upon Prince Turki's marriage. He eventually settled in an exclusive high-rise condominium complex called the Cricket Club in North Miami.

Miami police received reports that a female employee was being forced against her will to work seven days a week for little compensation; she was, in effect, a slave of Prince Turki. Bearing a search warrant, the police raided the two-floor apartment. Though the police found no slave, a violent brawl ensued. In a tape recording made by the police, a female, identified by the police as the prince's wife, is heard screaming, "I'll break your nose." At another point, the prince yells: "Nobody's leaving here. No! No! No! I have immunity. I challenge you." Afterward, Prince Turki, his wife, and his mother-in-law filed a $210-million lawsuit against the Dade County Police for violating their civil rights. State Attorney Janet Reno immediately filed a countersuit for violence committed against the police. The Saudi government then demanded that Turki be given diplomatic immunity.

At the request of the State Department, which was pressed by the Saudis, John West and Richard Gookin, social chief of protocol at the State Department, went down to Miami to iron out the difficulties and tell Prince Turki that the State Department was sympathetic to his problems. In a few days West reported to the State Department that Turki had been treated "outrageously."[30] "I came to the conclusion that a really basically erroneous charge had snowballed into what could be a terribly difficult and embarrassing international situation," West said.[31] Within a week Turki was granted diplomatic immunity despite the fact that he held no official title, the normal requirement for such diplomatic status. According to State Department documents, King Fahd had demanded this immunity.

West continued to inject himself into the dispute, though he told me that he had been invited down in April by the Miami chief of police to give a briefing to the police officers. On his third visit to Miami in two months, West ended up enraging the police, who had been fuming at the State Department's interference in a local police matter. After arriving in one of Prince Turki's limousines, driven by one of the prince's bodyguards, on his self-described "Good Samaritan" mission, West lectured the police that treating the Saudi prince "wrongly" could "affect forty thousand Americans in Saudi Arabia plus [have] possible international implications in terms of the overall relationship between the two countries."[32] Furthermore he told the police that he came to Miami as an "interested citizen," paying "my own expenses."[33]

Sitting in his plush Hilton Head office filled with Persian rugs and decorated with Middle East artifacts, I asked West in March 1984 whether he advised Saudi companies in his new business endeavors.

He said, "I do not have any clients who are Saudis that I advise about U.S. investments. I sort of stayed away from getting involved in Saudi companies or in Saudi government agencies. I have no connections there simply because I'd like to keep what credibility I have, and I don't like to be accused of being bought with Saudi money."

"So, you have no Saudi business?" I asked.

"No," he insisted. "A lot of my Saudi friends come over, and they ask me what I think about something. I tell them but I don't charge them." One of the recipients of his free advice is his friend Ghaith Pharoan, who is developing a golf course in Richmond Hill in Savannah, Georgia.

Ever since he left the diplomatic post, West has maintained that he has no direct business ties to Saudi Arabia. During the much heralded pro-AWACs press conference on September 22, 1981, at the Hyatt in downtown Washington, West (as well as three other former ambassadors to Saudi Arabia) was asked whether he had any business interests in Saudi Arabia. "No, I wish I did," he responded.[34]

When he flew to Miami to lecture the police on behalf of Prince Turki, he told Al Messerschmidt of the *Miami Herald* that he had intervened because "it is not nice to live in a hostile environment. That's the reason I wanted to make peace."[35]

But barely four months after he stepped down as Ambassador, West formed a consulting service, called West Advisory Services, later changed to the West Company, designed to assist American companies in getting Saudi business. West's clients are American companies that want to penetrate the Saudi market; West provides them with strategy, and more important, access to Saudi royal family, leaders, ministers, industrialists, and other businessmen. West's partner is David Burden, a former Citibank official who served in Saudi Arabia.

A recent addition to West's firm is Robert Walker, a former Treasury Department attaché who was stationed in the American embassy in Riyadh. One of the firm's first major initiatives was hosting a seminar in Hilton Head for officials of the $100-billion reserve Saudi Arabian Monetary Agency. West has organized business delegations, including one contingent of American bankers from the Southeast, on expeditions to Saudi Arabia.

West has also made trips to Saudi Arabia, most recently in February 1984, in his capacity as chairman of the American-Saudi Business Roundtable, a high-powered group he formed in 1981. Composed of executives from over fifty leading corporations, such as General Electric, Westinghouse, TWA, Pepsico, Marriott Hotels, and Chase Manhattan Bank, the Roundtable, which has offices in Washington, is designed to bring American businessmen in direct contact with Saudi officials. On the February 1984 trip, the Roundtable's delegation heard Crown Prince Abdullah demand that the U.S.–Israeli strategic cooperation agreement be immediately "renounced."[36] After the trip, West handed over the chairmanship of the Roundtable to his successor in Saudi Arabia, former Ambassador Robert G. Neumann, who was fired by Secretary of State Haig in 1981 for reportedly adopting an overly zealous pro-Saudi disposition.

Other American corporations also recognized the value of West's connections. In 1981 he was elected to the board of directors of Donaldson Lufkin & Jenrette (DLJ), a Manhattan-based brokerage and investment banking firm. At the time over 20 percent of DLJ stock was owned by Saudi businessman Suliman Olayan and Prince Khalid bin Abdullah, a brother of the Saudi king. (By late 1983, Saudi ownership of DLJ had increased to 23 percent.) In July West was asked to join the board of directors of the Whittaker Corporation, a hospital supply firm that earned more than a third of its fiscal 1983 revenues from Saudi Arabia. West, who was appointed head of Whittaker's international business affairs committee, was to be compensated at the rate of up to $40,000 a year for serving as a director and as chairman of the committee.

Besides his consulting service and board directorships, West also resumed his law practice in both Hilton Head and Camden, working with American companies that did business in Saudi Arabia. Finally, according to a source, West had been retained by one Saudi firm during late summer 1981 for $10,000 a month. According to this source who saw a copy of the check, a cover letter to West from an official of the Riyadh firm stated, "Here is your monthly retainer." The source did not know how long West had been retained by this firm.[37]

Besides the impact on public opinion he makes in his speeches and interviews, West has also ensured that his political views will be disseminated through his foundation. Formed in 1975 after he left the governor's office, the West Foundation was initially endowed with $50,000, raised by local friends and supporters, to fund the John C. West Chair in International Studies at the Citadel, West's alma mater. For five years the chair was the extent of the foundation's endeavors. At the beginning of 1981, according to IRS records, the net worth of the foundation was only $1,392. Suddenly, the foundation received a $500,000 contribution shortly after West retired as ambassador in March 1981. By the end of 1982, an additional $400,000, consisting of donations and interest earned on existing funds, had been added to the foundation for a total of over $900,000.

Since that time, the West Foundation has provided educational scholarships, endowed special chairs at South Carolina colleges, and funded cancer research. Another principal activity of the foundation has been to parcel out strings-attached grants to various educational and public policy institutions to fund symposia on the Middle East. The grant-making exercise produces a veritable public opinion domino reaction that pleases everyone: the foundation makes its grants; institutions receive much-needed funding; and the newspapers publicize the seminars. Who could object? After all, the West Foundation is offering academia new ideas. But the problem is that the West Foundation has been secretly funded by Saudi money. The recipients of the special politically restricted grants have become unwitting vehicles for the advancement of a foreign country's political agenda and the private vested interests of John West.

Soon after West left Saudi Arabia in 1981, a "Saudi citizen," according to a source intimately involved in the foundation, gave the West Foundation a donation of $500,000. Another Saudi who made a contribution was Prince Turki, the Saudi prince who received diplomatic immunity following his fracas with the Miami police. Turki apparently made his contribution as a gesture of gratitude to West for his intercession at the time.

When I asked West whether the foundation had received any Saudi funding, he indicated he was not sure: "There could be," he said, immediately adding, "I don't have anything to do with it [the foundation]." West advised me to speak to his brother who is chairman of the foundation. Actually, the foundation is officially run by his brother, S. J. West. And the foundation's only two directors are West's wife and son. West's longtime aide, J. M. Whitmire, is the treasurer. Both S. J. West and Whitmire refused to disclose, or even discuss, the contributions to the foundation.

Among the first recipients of a West Foundation strings-attached grant was the Southern Center for International Studies in Atlanta, a prominent nonprofit educational institution that has sponsored numerous symposia and seminars on international affairs. The center's board of trustees includes such illustrious names as Ruth S. Holmberg, chairman of the center and publisher of the *Chattanooga Times;* Cyrus Vance; Dean Rusk; Bert Lance; and Sam Nunn, the Georgia senator. West, who served as the center's first chairman in 1975, also sits on the board, as does Crawford Cook. In late 1981, West contacted the center and offered a "restricted grant" to set up a "series of colloquia focusing on Middle East issues." The center willingly accepted the offer and proceeded to organize a major conference.

On April 22 and 23, 1982, the center—under a $10,000

West Foundation grant, covering 43 percent of the $23,362 total cost—hosted a West Foundation–sponsored "colloquium on the Palestinian Question." Twenty-one academicians from nine surrounding southern states were flown to the conference and provided with hotel accommodations. Former Secretary of State Dean Rusk delivered the keynote address; the colloquium panel members consisted of two Israeli academicians, the Egyptian ambassador to Canada, two Palestinian professors, and a former high-ranking State Department official. No officials from the Israeli government attended, though they had been invited.

A West Foundation brochure published in 1983 proudly boasts that "this conference is believed to be the first instance in which Americans, Israelis, Arabs, including spokesmen for the Palestine Liberation Organization, had been engaged in a dialogue." [38]

The following year the West Foundation helped set up another Middle East conference. It provided a grant to the Citadel to present a "Conference on Gulf Security." The Southern Center for International Studies agreed to host the conference, sending out invitations to its members. On November 12, 1983, two hundred students, local residents, foreign policy specialists, and Arab government officials attended the all-day Saturday conference.

In spite of the discussion topic—Gulf security—some speakers could not resist promoting the PLO or attacking Israel. This was not surprising, given the makeup of the panel moderated by West: three State Department Arabists, one former National Security Council aide, two former high-ranking American military officials, Sheikh Ahmed Siraj, a minister of the Saudi embassy in Washington, and a former deputy foreign minister of the Saudi government. Siraj disregarded his prepared text; his impromptu comments revolved almost exclusively around his assertion that Middle East "instability" was "attributable to the creation of the state of Israel." [39] Not one of the invited speakers presented Israel's point of view.

Living the good life in Hilton Head, West continues to promote his political views. Arab government leaders, according to State and Commerce Department officials, hold West in extremely high regard. The net result to West has been quite profitable.

In many of his speeches, West is fond of saying, "Good business makes for good politics. And with poor politics, you will have poor business, or perhaps no business at all."[40] With an exemplary demonstration, West has proven his adage true beyond a shadow of a doubt.

The total transformation of the Honorable John C. West serves as testimony to the irresistible allure of the petrodollar.

In a neighboring southern state, however, the petrodollar's impact succeeded in capturing not just the allegiance of prominent individuals but the independence of an entire city as well.

19

THE ARABIAN SOUTH: BIRMINGHAM, SAUDI ARABIA

They [the Jews] will kill you with
a death of a thousand cuts.

*Former Congressman
Paul McCloskey, quoting
an unnamed friend*

To many people, Birmingham, Alabama, still conjures up memories of the white violence against blacks during the civil rights movement. It is not easy to forget the former police commissioner, Bull Connor, unleashing German shepherds against civil rights marchers and caring so little about the tragic deaths of four young black girls killed in the Sixteenth Street Baptist Church bombing. Few people are aware, however, that Birmingham corporations and other firms located in Alabama may very well hold the highest percentage of any state of $60 billion in American contracts with Middle East countries, and that as a result the dark side of petrodollar influence on the entire Birmingham community has been evidenced to an unparalleled degree.

Though the Birmingham firms' relationship with their Arab trading partners began quite innocently, it quickly assumed a life of its own. After almost a decade, the relationship culminated in a nasty episode in 1983 involving anti-Semitism and the illegal laundering of Saudi payments to a congressional candidate through a Birmingham corporation.

Built up after the Civil War on a mountain rich in coal and iron ore, Birmingham developed as a major southern industrial center. The unusual combination in Birmingham—which has a metropolitan population of 750,000—of raw materials for steel also led to the emergence of several of the South's—and the country's—largest construction, steel, and engineering firms.

Like other business communities across the United States, Birmingham sought to capitalize on the vast trade potential following the enrichment of the Arab oil producers. By 1976 Alabama firms had won a handful of contracts that were among the largest yet awarded by the Arab world. Blount, Inc., already considered Alabama's largest and best-known firm, landed a $120-million housing construction contract in Saudi Arabia. The Montgomery-based Blount is headed by Winton Blount, who was postmaster general under Richard Nixon. Blount's brother, W. Houston Blount, headed the Vulcan Materials Co., located in Birmingham and also one of the top firms in the state. A leading manufacturer of chemicals, secondary aluminum, and crushed stone, which is utilized in all types of basic construction, Vulcan received a five-year contract from Aramco to build and operate four stone-crushing plants and eight ready-mix concrete plants in Saudi Arabia.

Another Birmingham multinational, the Harbert Corporation, went to work in 1975 on a $52-million, seventy-mile water pipeline to transport fresh water from a desalination plant to Abu Dhabi. Though Harbert had built power plants, gas and oil facilities, and sewage plants in South America and Africa, this was the firm's first construction project in the Middle East. The contract—which Harbert undertook together with the Paul N. Howard Co. of Greensboro, North Carolina—also directed petrodollar contracts to other Birmingham companies. American Cast Iron Pipe Company, for example, received an order from Harbert for sixty-eight miles of pipe weighing 33,080 tons. The five ships that transported the pipe to Abu Dhabi constituted the largest single order for the port of Mobile.[1] Because of the earlier than anticipated completion of the freshwater pipeline, Harbert and

Howard were given another Abu Dhabi pipeline contract—this one for $47.5 million.

Within the next two years, the contours of the unique relationship between Birmingham and Saudi Arabia began to emerge. On Memorial Day 1978, twenty-five of the city's corporate executives, educational officials, and civic leaders met on top of Red Mountain at the Club, with its glorious view of Birmingham, to discuss ways to enhance business opportunities in Saudi Arabia for Alabama firms. According to a participant, an official from the Saudi embassy in Washington flew down to talk about the possibilities for trade and to thank the Birmingham business community for its "friendship." The Saudi official was so impressed with his hosts that he pledged $50,000 to build an adjunct to the Club and another $50,000 to local colleges.[2]

Participating at this gathering was the upper crust of Birmingham, the city's movers and shakers, including Houston Blount; David Vann, mayor of Birmingham; S. Richardson Hill, president of the University of Alabama in Birmingham; Neal Berte, president of Birmingham-Southern College; Dan Hendley, president of First National Bank of Birmingham; Ben Brown, Alabama vice-president for South Central Bell; Harry Brock, Jr., president of Central Bancshares of the South, Inc.; Alex Lacey, vice-president of Alabama Gas Corporation; Van Scott, board chairman of Brookwood Hospital; Emory Cunningham, president of *Southern Living* magazine; and Henry Graham, president of the Rotary Club of Birmingham.[3]

The next day, the group met again at the Downtown Club, where they heard guest speaker Miles Copeland, a retired CIA agent who had served many years in the Arab world and was now a consultant to American multinationals. Copeland, who was born in Birmingham, had been hired by twenty Alabama firms to help initiate direct contact with Saudi leaders. To this gathering Copeland explained the "terms of the Saudi-Birmingham friendship from a Saudi point of view."[4]

Much depended, Copeland said, on the outcome of the Egyptian-Israeli negotiations scheduled to take place at Camp

David in early September. Copeland vividly explained the link between politics and trade; one business executive commented to me that as a result of that meeting, "we were going to be the best damn friends the Saudis ever had." A special committee composed of business executives and educational officials were also formed. It was chaired by John E. Davis, Jr., head of Birmingham's largest architectural firm. Among those asked to serve on the Birmingham Middle East Business Development Committee were the president of the University of Alabama at Birmingham, S. Richardson Hill, and Houston Blount, the head of Vulcan.

That southern hospitality must have worked fast, for less than two months later, Copeland told Alabama businesses they could expect a whopping 4.5 billion dollars' worth of contracts from Middle East oil-exporting countries over the next three years, and possibly up to $50 billion by 1990. Speaking to the Rotary Club of Birmingham on August 15, 1978, Copeland said the reason that Alabama businesses were so successful was "because of the high regard held for Alabama firms already by Saudi Arabia." He pointed out that Birmingham was the only American city whose business leaders were directly in touch with the leaders of Saudi Arabia, a connection the ex-CIA operative had helped facilitate. "There is now a special relationship between Saudi Arabia and Alabama because of our past experience," he declared in a luncheon speech before the only Rotary Club in the United States that barred blacks from becoming members. Politics was very much on Copeland's mind. "Saudi Arabia has a policy totally consistent with our own. They want the same things," he said. And, he added, "We are particularly seeking out those Jewish Americans who have expressed embarrassment at threats which members of the Jewish community have leveled at Americans whose enthusiasm for Israel is less than the Jewish community thought it should be."[5]

The city of Birmingham was delighted by the prospect of the huge contracts with Saudi Arabia and other Middle East countries. The *Birmingham News* ran an editorial expressing its elation about the new multimillion-dollar trade prospects: "Copeland and others who have had a part in showcasing Alabama goods and services are to be congratulated. Their coup may be the biggest business story of the last quarter of the century for Alabama."

The following year Alabama's special relationship with Saudi Arabia took a giant leap forward. In April 1979 Saudi Prince Bandar, the Saudi defense attaché, and his diplomatic entourage from the Saudi embassy in Washington flew to Birmingham for a private and unannounced meeting with the Alabama governor and members of the Birmingham Middle East Business Development Committee. For both camps the meeting was considered an important coup. The Saudis valued opportunities to reach out directly to their natural American constituencies, the business community, and create political goodwill. The Alabamians cherished the chance to establish a direct pipeline to Saudi leaders, which, they hoped, would facilitate more contracts.

By 1980, Middle East business was booming for Alabama firms. According to Commerce Department records and State Department officials who served in the American embassy in Riyadh, Alabama businessmen made thousands of visits to Saudi Arabia. It was not out of the ordinary for some executives to have made seventy-five trips to the Arab world over the course of six years. The disproportionate frequency of their trips was so great and their presence in Saudi Arabia so conspicuous that they earned the nickname "Alabama Mafia" from U.S. Ambassador John West. No less than fifty companies from that state signed contracts or subcontracts with the kingdom and other Middle East countries, ranging anywhere from $25,000 to $250 million. Over a dozen of these companies established offices in the Persian Gulf.

Suddenly, in early May 1980, a major flap in U.S.–Saudi relations sent shock waves throughout Alabama firms and other American businesses involved in Saudi Arabia. The Public Broadcasting System reiterated its intention to broadcast as scheduled *Death of a Princess*. A furor erupted as Saudi Arabia demanded that the film not be shown on American TV. In Birmingham, business went immediately to work. (For a full account of Birmingham's involvement in the *Death of a Princess* episode, see Chapter 9.)

Bill Harbert, then executive vice president of the Harbert Corporation, called Alabama Public TV to voice his strong objections. Other expressions of outrage were voiced by other members of Birmingham's corporate and civic elite, such as officials of Blount Inc. Alabama businessmen made over three dozen telephone calls protesting the film. On May 11, an official of the five-member Alabama Educational Television Network announced it would not air the film.

Yet, even after the film had been suppressed on Alabama TV, Alabama corporations were still fuming. On Sunday, May 18, 1980, six days after Alabama Public TV had withdrawn the film, the Harbert Corporation took out a full-page advertisement in the *Birmingham News* asserting in a headline: "The people who shout 'Death to America' and the people who made the film 'Death of a Princess' have a lot in common. Harbert Corporation refuses to support either." Condemning the American government's refusal to suppress the film on public TV, Harbert stated: "Saudi Arabia is often described as a 'moderate Arab state'. As moderates, they ask our government to help assure that our Public Broadcasting System not be allowed to falsely portray their culture, religion, and society. We think it a shame that our government didn't honor that request. It would be a lot better world if we didn't have to listen to shouts of 'Death to America' and if our friends around the world didn't have to wonder about a nation which would allow 'Death of a Princess' to be aired on Public Television as fact."

Seven months later, in December 1980, Alabama was assured that the Saudis still held the state in good grace. The Blount Corp., whose officials had actively lobbied against *Death of a Princess*, announced that the firm had received, after three years of bidding and negotiation, the "largest single international construction contract ever signed."[6] Together with a

French concern, Blount successfully landed a $1.7-billion contract for the construction of the University of Riyadh in the Saudi capital—a five-square-mile university complex. The plans for the institution called for building ten colleges, dormitories for twenty thousand students and their families, research institutions, parks, auditoriums, a parking lot for seven thousand cars, a library with a six-thousand-student capacity, a nine-thousand-person-capacity mosque, five dining halls with eight thousand seats, research laboratories, a sports stadium, parks, and even a mass transit system. As the major contractor, Blount was in a position to throw substantial business to other firms in the state.

The political link to Arab governments began to manifest itself in other ways. The First National Bank of Birmingham, a leading bank, helped negotiate a performance guarantee in 1981 with the Syrian government for a $19-million pipeline contract project with the U.S. Pipe and Foundry Company of Birmingham. In a quarterly brochure (touting the contract as well as economic development in Syria) published by the bank's international department, a map of the Middle East was included. The name "Palestine" appeared in place of "Israel." Later, the bank apologized for using "outdated material" and issued a corrected brochure with a new map naming "Israel."[7]

But the benchmark test of the special Birmingham-Saudi relationship occurred in early March 1981. President Reagan announced that the United States would sell to Saudi Arabia offensive equipment for their F-15 fighters and five supersophisticated, advanced AWACs surveillance planes. Congressional opposition to the sale snowballed.

For Alabama businesses, however, the prospect of the U.S. government denying a Saudi request sent shock waves through Birmingham's business community. In July, Winton Blount addressed the Birmingham Rotary Club where, wrote the *Birmingham Post-Herald*, he told his audience that "America's national interests rated more with Saudi Arabia than with Israel."[8] "The Saudis want to be American friends," he added. "They despair when we talk about whether we'll sell them AWACs." In a news conference after a speech, Blount displayed a copy of a check made out for the sum of $343,373,480 that he received as a partial payment for the construction of the University of Riyadh.

By late October, with congressional opposition to the sale of AWACs still mounting, Alabama businesses were deeply worried. So, on October 22, four planes carrying a total of twenty-seven Alabama business executives, including officials of Blount and Harbert, flew from Birmingham to Washington to demand that Senator Howell Heflin support the AWACs sale to Saudi Arabia. In a stormy meeting, according to congressional sources, the senator was told in unambiguous terms that he should not plan on gaining reelection if he voted against the sale. Resentful of this heavy-handed pressure, Heflin voted against the weapons package. The sale squeezed through, however, and Alabama businessmen breathed a collective sigh of relief.

After these political triumphs, Alabama businessmen sought new ways to engender Saudi goodwill. Informed of Saudi Arabia's desire to acquire Western recognition for their newly created educational institutions, the Alabama businessmen realized they could kill two birds with one stone: please Saudi officials and help their own educational establishment, the University of Alabama.

The University of Alabama system has three main campuses, with a total of 34,000 students, 3,000 teachers, and a $490-million annual budget; Birmingham (UAB), in the center of the state, is the largest of the three campuses. The other two are in the smaller towns of Tuscaloosa and Huntsville. S. Richardson Hill, the president of the Birmingham division—with 20,000 students—was quite familiar with the dimensions of Birmingham-Saudi trade as he had served, alongside Houston Blount, on the Birmingham Middle East Business Development Committee. Hill had also been elected to the board of directors of the Vulcan Materials Company, which had extensive ties to Saudi Arabia. In addition, Winton Blount was both a trustee of and an influential fund-raiser for the University of Alabama.

Like most major universities, the University of Alabama system was searching for new sources of funding. In particular, the university—due in part to President Hill—had for years been trying to cash in on the lucrative Middle East petrodollar market. But according to university sources, a problem in the past had been to devise legitimate bilateral programs that, amid the zeal to find money, would not confer unacceptable and undeserving respect on brutal regimes.

In early 1977, the university had considered setting up an exchange program with Libya but canceled it after members of the faculty protested. Tentative plans called for UAB to become aligned with Al-Fatah University in Tripoli—named after Muammer el-Qaddafi's coup d'etat. The Libyans were to pay for Birmingham medical faculty to teach in Libya and for Libyan students to attend school in Birmingham. Most important, the UAB favorably considered allowing Libya—a member of OPEC that facilitates and sponsors international terrorism—to set up a Center for Islamic Studies for which the university, according to UAB officials, would have received a princely sum. S. Richardson Hill traveled in February 1977 to Libya to discuss preliminary details of the arrangement. (He was appointed president of UAB while in Libya.)

Upon his return from the ten-day trip, Hill spoke glowingly of his hosts. Quoted in an official university publication, Hill said he came away "favorably impressed" with Libya: "They are making great progress with meeting the needs of the people."[9] Faculty indignation forced the university to cancel any further consideration of the program. In 1979, an exchange program between Iraq and the UAB School of Dentistry developed, but it is not a popular program.

The propitious moment for solving the needs of the university, the Alabama business community, and the Saudis arrived in February 1982. At that time, H. R. (Terry) Vermilye, the senior vice-president of Birmingham Trust National Bank, traveled to Saudi Arabia as part of a contingent of six bankers from the Southeast led by John West. On this trip West arranged for members of his group to meet Mansour I. Al-Turki, newly appointed president of the University of Riyadh.

Vermilye met with Al-Turki on February 16. At that meeting the Saudi University president, eager to expand the connections of his relatively new university, expressed his "interest in establishing bilateral relationships with institutions of higher learning" in the United States.[10] Upon returning to Alabama in March, Vermilye immediately contacted the University of Alabama in Birmingham president to inform him of his discussion with Al-Turki. In a letter to Hill, Vermilye wrote, "I am writing now because it occurred to me that the University of Alabama in Birmingham might wish to explore such relationship—perhaps in some area of medicine." Vermilye then pointedly reminded Hill of the fact that "a $1.7-billion contract for the construction of the University of Riyadh campus was recently signed with Blount International of Montgomery."

On March 29, four days after he received Vermilye's letter, Hill wrote to Al-Turki, expressing his hope of establishing a "Riyadh University–UAB relationship which would be mutually beneficial to both of our institutions." In his letter, Hill reminded Al-Turki that Blount International was building Riyadh University, that Winton Blount was a member of the board of trustees of the University, and that he (Hill) was a member of the board of directors of Vulcan Materials of Birmingham, a firm "which also has business interests in Saudi Arabia." Hill reported on these developments to an ad hoc University Middle East Committee, which he created immediately after Vermilye's trip. Composed of six faculty members and administrators from UAB, the committee convened for the first time in April 1982 to discuss links with Saudi Arabia and other connections to the Middle East, including one with Egyptian schools.

Months later, another separate group within the university system began to push for mutual university programs with Saudi Arabia. In March 1982 the University of Alabama system announced the appointment of a new chancellor, Thomas Bartlett, effective August 1. Bartlett, a former president of Colgate University and the American University of Cairo—his wife still represents the latter institution in Washington—was heavily promoted by Winton Blount, according to university sources.

After assuming his duties in Tuscaloosa, the fifty-one-year-old Bartlett, at the suggestion of Houston Blount, set up the Ad Hoc Committee to Consider Joint Programs with Saudi Arabian Universities. The group first met on the afternoon of October 21, 1982, at the chancellor's home in Pinehurst, Alabama. Over coffee and cake, Bartlett told the group of two deans and four professors, according to minutes obtained of the meeting, of trustee Blount's "interest in introducing Mr. M. Al-Turki to the University of Alabama community and of his own [Bartlett's] interest in promoting a beneficial relationship with one or more Saudi universities and various branches of our System."

Chancellor Bartlett's group was careful to consider the culture gap between Americans and Saudis and to emphasize the advantages of the southern state and the University of Alabama for any U.S.–Saudi exchange. Bartlett noted that Alabama was attractive because of "the strong religious background of the state and the past hospitality shown to Middle Easterners living here." One committee member suggested the university "make provisions for a mosque should we have significant numbers of Saudis on campus." In addition, university officials "must take care" to emphasize that the Saudis will get "more direct attention" and "more realistic and practical programs" than at the prestigious eastern universities such as Harvard, M.I.T., and Princeton. "The University of Alabama," noted one committee member, "is tied to reality, not an isolated ivory tower."

The short-term programs contemplated at the ad hoc committee meeting included faculty exchanges, special health care programs, engineering programs, and business seminars. Depending on the initial success of the collaboration, according to internal UAB records, a major "Middle Eastern Institute" was to be established within the University of Alabama system. One member of the ad hoc committee specifically pointed out that the university should display the appropriate sensitivity to "Saudi sensibilities": "We should recognize the Saudis' natural desire that people everywhere know more about their history and culture and, in particular, more about Islam as one of the major religions of the world."

During the remainder of 1982 and in the early months of 1983 the relationship between King Saud University (changed from Riyadh University) and the University of Alabama system progressed unevenly, as faculty members got bogged down in bureaucratic negotiations over programs. In biomedical fields, UAB was successful in interesting its Saudi counterpart in bilateral programs. But in other areas, UAB did not find academic fields for joint ventures. Still, President Hill, the UAB Center for International Programs, Chancellor Bartlett, the two ad hoc Middle East committees, and various Alabama businessmen kept meeting to devise new programs and services that the Saudis would accept. Representatives of the school were prepared to travel if necessary to Saudi Arabia to find out what exactly the Saudis wanted.

While university officials busily sought Saudi links, UAB's Hill was approached in October 1982 with an offer that would gratify Alabama's business community, the university's board of trustees, and Saudi authorities. An official of the American-Arab Affairs Council (AAAC), a new pro-Arab organization based in Washington, asked Hill if the university would be interested in sponsoring a conference called U.S. Economic and Political Challenges in the Arab World. According to an AAAC official, the organization had decided to hold the conference in Birmingham for two purposes: (1) to capitalize on the extensive corporate connections with Saudi Arabia, and (2) to avoid the scrutiny of the "Eastern" press.

AAAC's political disposition is clear. Most of its publications, interviews, and special reports focus on Israel as the exclusive source of repression, instability, and political upheaval in the Middle East. Although the AAAC's politics are emphatically one-sided, it is not difficult to see that the group has its own political agenda. At a faculty meeting one professor questioned the propriety of the university's sponsoring such a conference. However, a university official expressed the belief that "this [conference] will provide a worthwhile opportunity for discussion of significant international issues, to involve a wide variety of people in Alabama, especially from the private sector."[11] He added that AAAC "has established its credibility as a responsible agency engaged in promoting serious discussion and consideration of important issues affecting relationships between the U.S. and Arab countries."

Hill immediately offered official institutional sponsorship. The UAB president also agreed to serve on AAAC's planning committee and to recruit prominent businessmen to fund and attend the gathering. Chancellor Bartlett also endorsed the conference, agreeing to help promote the event and to recruit corporate participants, if necessary. In December 1982, a month before the conference was publicly announced, the university's decision was presented as a fait accompli to the UAB Middle East Committee, although two committee members expressed concern that the university might become wrongly involved in politics.

The university administrators were not, however, acting in a vacuum. According to Birmingham sources, Vulcan Materials, where Hill serves as a director, was also contacted by AAAC officials in late 1982. Vulcan officials responded eagerly to the request, volunteering to help organize the conference and to contribute $1,000 toward AAAC's expenses. (The following year, Donald Rumsfeld—the former secretary of defense who was selected by President Reagan in October 1983 to serve as the special Middle East envoy—was elected to Vulcan's board of directors.) Other corporations that subsequently agreed to make tax-deductible contributions of $1,000 included South Trust Bank, the holding company for Birmingham Trust National Bank, whose vice-president, Vermilye, helped initiate the UAB–King Saud University links; Sonat Incorporated, a huge multibillion-dollar Birmingham-based natural gas firm; Kellogg-Rust Incorporated, a multinational engineering-construction firm with offices in Birmingham, which was the largest American recipient in 1982 of Saudi contracts—over $4.5 billion; P. E. LaMoreaux and Associates, a Tuscaloosa-based engineering and consulting firm involved in setting up resource programs in Syria and Egypt; Clow International, a Birmingham-based firm that does tens of millions of dollars in business in manufacturing sewage treatment facilities for Saudi palaces and military bases; and Harbert International (a subsidiary of the Harbert Corporation).

Finally, the Alabama Small Business Development Consortium, sensing an opportunity to gain some exposure and business, also agreed to host the conference. As an official agency of the state of Alabama, funded substantially by the federal government's Small Business Administration, the consortium's sponsorship meant that the U.S. government had indirectly facilitated the gathering.

And so, bearing the academic imprimatur, AAAC formally announced in January 1983 that on March 3 and 4, AAAC "in association with the University of Alabama at Birmingham, the Center for International Programs at UAB, and the Alabama Small Business Development Consortium would cosponsor a major regional conference at the Birmingham Hyatt Hotel." According to the press release, the parley was to "focus on investment and business opportunities in the United States for Arab businessmen, and business opportunities in the Middle East for Americans." It would also include "a thorough consideration of the contemporary political climate and its implications for U.S.–Arab business cooperation." Invitations were sent to members of local chambers of commerce and business groups in southeastern states.

On Thursday evening, March 3, a reception was held in the Polaris room on the seventeenth floor of the Birmingham Hyatt. With bourbon flowing quite freely, the next day's speakers, University of Alabama administrators, Arab diplomats, participants, and Birmingham's corporate elite mingled at a party that seemed like a reunion of longtime friends. The next morning the conference commenced formally. Before entering the first-floor hotel lecture hall, conference participants were given special packets of materials by the organizers containing an agenda emblazoned with the university's name. Included in the packet were an interview with PLO adviser Khalid Hassan, a synopsis of the Middle East conflict that focused on "Zionist aggression" from 1948 to the present, and other informational items that distinctly accentuated the "political challenges" for the United States and the Arab world over the economic ones.

The majority of the 150 registrants were exporters, manufacturers, retailers, bankers, and consultants who conducted business with Middle Eastern countries or who sought to gain access to what was portrayed to them as the bottomless petro-dollar market. As I talked with numerous registrants during and after the conference, I discovered that many had never been exposed to the political dimensions of the Middle East conflict save for the scant information to be gleaned from the local media, and, though many who attended came from Alabama, a substantial number also came from Georgia, Texas, North Carolina, and Florida. Also participating were Arab embassy officials from Washington, Persian Gulf investors, a diplomatic official from Quebec, and representatives of several major investment banks and corporations such as Shearson/American Express, E. F. Hutton, and Northrop. The gathering also included over a dozen officials of the University of Alabama system, who attended in their official capacities, including President Hill and Chancellor Bartlett. Both sat in the back of the conference room throughout the entire day's program.[12]

James Akins, former ambassador to Saudi Arabia, delivered the opening address. The tenor of his speech was set by his analysis in his introductory comments, of why Truman recognized Israel: "In 1947 the State and War Departments argued strongly and convincingly against American support for the partition of Palestine. They pointed out cogently and accurately that the U.S. had very real interests in the Arab world. . . . Truman settled this issue at that time and apparently forever with his remarks, 'I understand what you are saying, gentlemen, but I have no Arab voters in my constituency.' "

Akins also defended the 1973 Arab oil embargo against the United States because of the "massive military and economic assistance" provided by the United States to Israel during the war. He defended Libyan dictator Muammar el-Qaddafi. Compared to Israel, the former diplomat asserted, the activities of Qaddafi have been benign. "What has Qaddafi done to hurt America's interests? Nothing. His relations with the oil companies have been better than almost any other country."

In the wake of the Israeli invasion of Lebanon, Akins asked rhetorically, "What did the Arabs do? Nothing." Akins added that cease-fire was "scrupulously honored by the Palestinians" until broken by the Israelis. "These attacks [by PLO on Israeli territory] have been relatively minimal for the past twenty-five years. The number of Israelis killed by Palestinians is one–one hundredth the number of Arabs killed by Israelis during that time."

Akins lamented the fact that "when the U.S. changed its laws against the Arab boycott, most Arab governments changed their laws to comply. He noted despairingly that "no Arab countries withdrew funds from American banks" after the massacre of Palestinians in September 1982. Expressing his exasperation, the former ambassador summed up his speech—in a somewhat ironic role for a former official of the United States— by criticizing Arab governments for not censuring the United States: "We have gotten everything we wanted from them after taking a policy quite obviously anti-Arab. Every Arab government listens to the U.S. and every Arab has swallowed the humiliation with no response to the U.S."

Ironically, the speaker who followed Akins, Harry Martin, president of the UBAF Arab American Bank, felt compelled to disagree with Akins about the impact of political turbulence on Arab investment in the United States. "When tension develops in the Middle East," Martin said as he rose to follow up Akins's response to a participant's question, "more money flows into the U.S."

The next major speaker was former Congressman Paul (Pete) McCloskey, Jr., who represented the seventeenth and later the twelfth district of California in the House of Representatives from 1967 through 1982 when he gave up his seat to make a bid for the Senate. McCloskey is now a partner in the San Francisco law firm, Brobeck Phleger & Harrison. As a maverick Republican, McCloskey, a leader of the anti–Vietnam War movement, even briefly challenged Nixon for the Republican presidential nomination in 1972. A charismatic self-styled progressive, the fifty-five-year-old legislator has assiduously built his reputation as a fighter for civil rights and as a courageous opponent of the Vietnam War. For example, two weeks after the Birmingham conference, McCloskey delivered the keynote address at a congressional symposium in Washington honoring the late liberal Democratic New York congressman and anti–Vietnam War activist and civil rights leader, Allard K. Lowenstein. In that speech, McCloskey spoke of the need to continue fighting for civil rights and "against anti-Semitism."

But before his Birmingham audience, the tone and substance of McCloskey's remarks were remarkably different. He began his speech by informing his audience of his new career and soliciting them for petrodollar business:

> *I am in the same business as many of you are now. My clients would like to consider investing in the Arab world, they would like to consider attracting Arab investment into their business and real estate. They would like to get contracts with Arab countries for the construction in Arab countries or elsewhere of major projects.*

McCloskey then predicted that Arab investment and trade in the United States will be predicated upon the willingness of American businessmen to lobby for the Arab nations. He minced few words:

> *I suspect that every Arab businessman and certainly every Arab government with which many of you may deal in the next year may very well insist that as a condition for their investment with you or their contract with you that you begin to play a part in negotiating with and lobbying with the Congress of the United States. . . . Because it is the Congress today which denies to the President of the United States the ability to advance peace in the Arab world which is consistent with the Arab interests.*

Across the Hyatt reception room, McCloskey's words elicited total silence as he explained why it was incumbent upon the business community—if they wanted to "promote investment opportunity"—to become active lobbyists on Capitol Hill: "Because the Congress is literally terrified of the most powerful lobby in the United States which is called the Jewish community—or AIPAC, which is its head." He continued, "For the first time in history, you have Congress operating in support of the Jewish community in America, denying the President of the United States the ability to conduct his own peace plan." McCloskey, the social crusader, had found a new way to attract business.

As he spoke, without notes, his words seemed to flow naturally as he pandered to the crowd of southern businessmen and Arab investors:

> *What really concerns a politician is the view of the small single-interest, dedicated group which has religious zeal in promoting its view on that issue—a group which may constitute only two percent of the American people but which is highly organized, highly motivated, contributes inordinate financial support, works hard in elections and votes all because of a*

candidate's view on a single issue—that is the polit-
ical view that controls America today. That kind of
group literally controls foreign policy in the Middle
East today and in the Congress.

McCloskey advocated immediately cutting off aid to Israel unless Israel signed the nuclear non-proliferation treaty and removed its troops from Lebanon by "tomorrow." And, he added,
"the only reason that the President doesn't [cut off aid to Israel]
is knowing that the Congress is approached by a lobby on only
one side of the issue." Omitting the two fiercest political battles
over American-Middle East policy—the F-15s in 1978 and the
AWACs in 1981—that the Jewish lobby lost in Congress,
McCloskey completed his diabolical portrait: "Congressmen will
never have the guts to take the position in opposition to the Jewish community of America."

Leaning over the microphone and speaking in a subdued
voice, McCloskey warned of retributions from the Jewish community:

I would like to say one final thing. Those of you who
have the courage to suggest that Israel is wrong may
face economic sanctions in this country. They will be
hidden sanctions. A friend of mine—a president of one
of the major corporations in America, my roommate;
we enlisted in the Marine Corps together many years
ago—told me when I made the statement I did, that
if we were to have peace in the Middle East, we had
to respectfully disagree with our Jewish constituents.
He said, "Pete, they will kill you with a death of a
thousand cuts." Whatever Arab Americans are, they
haven't yet achieved prominence in the television net-
works, the radio industry, the news media [as the
Jewish community has]. They have not yet reached
the financial affluence that the Jewish community has
achieved.

McCloskey insisted that his comments were made out of "deep respect" for the Jewish community, adding that "they work hard, they are family oriented, they succeed. You can't find a profession, the academic world, the law, business, the arts, where there aren't fine Jewish leaders." He professed that he was concerned about anti-Semitism: "The last thing we want to see is any rise of anti-Semitism in America." But in the same breath he charged that the national loyalty of Jews is toward Israel and suggested that a Jewish cabal exists:

> But the way to stop anti-Semitism is to reach out to those Jewish leaders and say: You are the people who can turn Begin around. But you have got to tell Mr. Begin that he can no longer rely on your lobbying secretly with the U.S. Congress to support Israel right or wrong. And I say this, the Jewish community operates secretly in this respect [emphasis added].

McCloskey had attacked the lobbying of American Jews as illegitimate, traitorous, and detrimental to American interests. But he urged American businessmen to do the same kind of lobbying. After then citing three examples of "economic sanctions" to which he claimed to have been subjected by Jews, he concluded his speech with one final attack:

> Hopefully we will never see a rise of anti-Semitism again. But unless American businessmen will lobby with individual congressmen and senators, the Congress will continue to do what it quietly did last year, give more money instead of less to Israel. And until this matter can be discussed publicly in the business community without fear that some Jewish customer will withdraw his advertising from a TV station or take away his bank deposits, I suspect that Congress will continue to do what is very seriously jeopardizing the chance of peace.[13]

The audience burst into applause.

At noon the conference adjourned for lunch where Birmingham Mayor Richard Arrington, who had just returned from a trip to Jordan, greeted the participants. "These kind of conferences," Arrington said, "lead to better understanding" between the United States and the Arab countries. He was followed by another luncheon speaker, Ghanim Al-Mazrui, secretary general of the Abu Dhabi Investment Authority. Pointing out that "one hundred twenty states recognize the PLO, but only twenty recognize Israel," Mazrui declared that "pressure groups are diverting American foreign policy from its best course."

At the end of the day's conference—during which former Ambassador John West delivered one of his stock speeches on the Middle East—H. Brandt Ayres, publisher of the *Anniston Star,* a daily Alabama newspaper, rose to deliver the "summary of discussion." But the liberal commentator, whose own presence at this conference had raised questions by some observers in Birmingham, was clearly shaken by what he had heard throughout the day.

When he stood up, the tall, bespectacled southerner said he felt compelled to make just one comment. Ayres warned of the dangers attending to singling out "the Jewish conspiracy." Flanked to his right by West, who sank his head into his folded arms as Ayres spoke, he told the crowd, "If we attempt to make the [Palestinian] issue come clear and exert our persuasive powers as strongly as required to get a solution to the problem at the cost of indifference or enmity to the Jewish people or Jewish state, then we pay too high a cost." Ayres warned of painting a picture "of six million foreign agents who can and do manipulate a nation of 230 million because they expertly know the pressure points of our society."

Later in the evening, a banquet was held at the Relay House. In attendance were many conference participants and top executives from Alabama's leading corporations. UAB President Hill even purchased a table of tickets. The banquet address was given by Abdullah Alireza, head of the Persian Gulf conglomerate.

Alireza, who also serves on the boards of directors of several banks in the United States, condemned "the expansionism" of Israel, and the "Zionist lobby." At Alireza's side was Roy W. Gilbert, Jr., chairman of the South Trust Bank, who was voted 1981 Man of the Year in Birmingham. Alireza urged American and Alabama businessmen to lobby for "our best interests." Wrapping up the conference, Gilbert dwelled on the attractiveness of Alabama as a "good place for Arab investment."

When I asked why Birmingham Trust National Bank played such a supportive role in facilitating the conference, H. R. Vermilye, the bank's vice-president, said, "I did it because I thought it was in the best interests of our bank, city, and country." He continued, "There is an inadequate appreciation of the Arab point of view. . . . We believe as members of the business community that it is in our firm commercial and geopolitical interests to develop a multifaceted relationship with the Arab countries of the Middle East." He added, "If we take actions that in their [the Saudis'] view are unreasonable, imbalanced, or opposed to the interests of their country, they will be less inclined to do business with our country." As for direct pressure generated by the Saudis and other Arab oil producers, Vermilye said, "I've never had it thrown up in the form of blackmail. But enlightened self-interest? Yes." [14]

Enlightened self-interest may also explain the involvement of officials of a Birmingham corporation in a fund-raising episode that involved the possible laundering of Saudi money.

On March 18, 1983, according to Federal Election Commission (FEC) records, Paul Findley, a former Republican Congressman with pro-Arab sympathies who had just been narrowly defeated in his 1982 campaign, received twenty checks, each for $1,000, as political contributions to his political action committee. All the checks were recorded as having been received on the same day—March 18, 1983—and half of the donors shared a common address: a post office box in Birmingham. The other half listed addresses from states as far apart as Texas and New York. To a casual reader of the FEC records, it would have seemed that Findley had been honored with a post-election fund-raising event in Birmingham.

In fact, though, the checks were written in 1982, presumably in time to have been used in Findley's reelection campaign. And the fund-raising occurred many thousands of miles from Birmingham; at the time, all the contributors lived and worked in Saudi Arabia. And the company for which they worked in Saudi Arabia or by which they had been recruited to work in Saudi Arabia was the Vulcan Materials Company, which had played a prominent role in hosting the AAAC conference.

One of the donors, Jack Callaway, who provided details about the fund-raising, admitted he received a cash reimbursement for $1,000 several weeks after making his gift to Findley. He added that he had never heard of Findley until approached by an American who works as a middleman for a Saudi businessman. The American middleman took possession of the check from this donor and from others and forwarded them to Findley. Another American executive also admitted that he was offered a $1,000 reimbursement for making a contribution to Findley.

According to Callaway and the other source, the fund-raising effort was carried out at the direction of Khalid Ali Al Turki, a joint partner with Vulcan in several Saudi ventures that have generated $100 million in annual revenues in recent years. Al Turki, whose personal wealth is thought to be well over $100 million, represents numerous foreign companies in Saudi Arabia such as Lockheed Missiles and Space Co. (Sunnyvale, California) and Nippon Electric Corporation (Japan).

Al Turki and Vulcan set up a joint venture, Tradco-Vulcan, in 1976. The joint venture was set up through Al Turki's holding company, the Trading and Development Company, known as Tradco. Described by those who work for him as urbane and publicity-shy, Al Turki is close to the Saudi royal family and has been rumored at times to be a candidate for a ministerial appointment.

Al Turki's interest in raising campaign funds surfaced at

about the time an unusual plea was published on October 22, 1982, in *Al-Jazira*, the Saudi newspaper. Although published in the Arabic language, the article was intended for the community of American businessmen living in or conducting business with the oil-rich country.

As translated and distributed by the Foreign Broadcast Information Service (FBIS), a part of the Central Intelligence Agency that monitors all foreign media, the article described Findley's political plight and implored U.S. companies doing business in the Middle East to help the pro-Arab Congressman win his election: "[W]e know that the congressional game in that country is not distant from the fingers of Zionism and the influence of its vast and skillful media. . . . All that Findley now needs is $150,000 to $250,000 to win. Is this amount too much for companies and establishments to contribute through political action committees?" The article went on to suggest that companies responding to the request could expect to "reap manifold benefit" from their generosity.

Al Turki did not personally engage in the fund-raising effort, Callaway and others said, but left that task to Gorton DeMond, a U.S. citizen who is vice-president of Al Turki's holding company. DeMond, who was recruited to work in Saudi Arabia in 1976 by Vulcan, helps manage Al Turki's interests in the Saudi businessman's partnerships with Vulcan.

Callaway, a Birmingham native who had been recruited by Vulcan to manage one of Al Turki's other companies, said he was told by DeMond: "Al Turki would like you to make this contribution. It won't cost you anything out of pocket." A forty-nine-year-old businessman who went to Saudi Arabia to make some quick money, Callaway said he wrote a $1,000 check, handed it to DeMond, and several weeks later received an envelope containing $1,000 in cash from the accountant at Al Turki's holding company. Another Vulcan-connected employee who was in Saudi Arabia at the time also said DeMond had offered to reimburse him for a $1,000 contribution, but the employee refused to make the donation. Both Callaway and the other source said they had heard that Al Turki had intended to raise $40,000 to $50,000 for Findley.

Callaway said he had never heard of Findley before being asked to make a contribution. He did so, he acknowledged, because it was not going to cost him anything. "I didn't know what he was running for," the executive said in a November 1983 interview. Callaway's willingness to talk to me and his honesty stood in contrast to the responses of other Vulcan employees. Most would not agree to talk at all. And of the seven who agreed to be interviewed, all denied getting any reimbursements for their contributions. But several admitted to being similarly unfamiliar with Findley until DeMond approached them for money. When I asked one donor why he contributed $1,000 on the spur of the moment to a candidate he had never heard of, he said: "I thought it would be good for the company and the long stay." The annual salaries of most of the contributors ranged from $40,000 to $70,000.

DeMond, reached by telephone in Saudi Arabia, admitted he approached "several" Americans for contributions but denied he offered any reimbursements for their gifts. Asked who had asked him to raise funds for Findley, DeMond claimed he was approached "by a group of people in the United States" who asked him to "see if people [in Saudi Arabia] were interested in contributing." He refused to say where the request came from, except to say "they were not directly linked to Vulcan or Findley—they were not connected to anyone." A Vulcan executive in Birmingham denied in an interview with me the company had any knowledge of the fund-raising for Findley, despite the fact that top officials of Vulcan's Saudi operations were listed as $1,000 contributors and that most of the donors still work for Vulcan either here or in Saudi Arabia. When asked about the donations, Findley said they had come unsolicited in the mail and that he had never heard of the name Tradco-Vulcan before I mentioned it to him.

The Federal Election Campaign Act bars contributions to candidates for federal office by foreign nationals and prohibits the use of other people to mask the true identity of the contributor. Willful violation of either of these laws involving $2,000 or more can result in maximum penalties of a $25,000 fine or one year imprisonment.[15]

In April 1984 the saga of the special Birmingham-Arab relationship continued to unfold. Mayor Arrington traveled to the United Arab Emirates and to Kuwait—at those countries' expense—in an effort to attract Arab investment. Arrington met with Arab government and banking officials. "We ought not pass up prospects of investments from any reputable source," said Arrington upon his return. The mayor also admitted to reporters the motivation of his hosts: "Good business is good politics, that's what they said to me. I think that they feel the more linkages they make through investments in this country, the more friends they build and undoubtedly they hope there's going to be some benefits accrued in terms of America's foreign policy as it relates to those countries."[16]

Birmingham was not the only southern city where the petrodollar contagion spread. In early March 1983, about the same time that the Birmingham conference was taking place, Mayor Andrew Young of Atlanta initiated the first step toward building the financial bridges between the Arabian Peninsula and the Peach State, traveling to the United Arab Emirates and Saudi Arabia. Everywhere he went, the former U.N. representative was afforded a welcome befitting a returning hero—in fond memory of his meeting with a PLO representative in 1979.

Throughout his sojourn to the Gulf, he was given first-class treatment and VIP benefits such as $8,000-a-pound incense. In Riyadh, he occupied a seven-room hotel suite replete with three baths and a manservant. Donning the traditional long white flowing Arabic robes in his hotel and placing a kaffeiyeh around his head, Young proclaimed himself "Sheikh Andy."[17]

Young's airfare and hotel bill in Abu Dhabi were paid for by the UAE government. In another emirate, Dubai, the National Bank of Georgia, now owned by Saudi billionaire Ghaith Pharoan who purchased it from Bert Lance, picked up the hotel tab. In the eastern provinces of Saudi Arabia, Calocerinos & Spina, a Syracuse, New York–based engineering consulting firm that had recently opened an Atlanta office, paid the mayor's expenses. A dinner in Young's honor was held in Saudi Arabia, at which his friend Abdullah Bishara—whose New York City home was the site of Young's secret meeting with the PLO official—was one of the guests.

Throughout his visit, Young invited business officials to attend a conference on Saudi Arabian trade that was to be held in Atlanta in May 1983. Sponsored by the U.S. Chamber of Commerce and the *Saudi Gazette,* an English-language business daily based in Jidda and owned by OKAZ Communications Company, the conference was to focus on opportunities outside petroleum-related areas, according to a conference spokesman.

On May 9 and 10, 1983, three hundred American businessmen and bankers gathered at the Hyatt Regency for the Saudi-American conference.[18] For $1,500, a corporate guest received three admission tickets, conference acknowledgment in a special brochure, and a table for eight at a banquet with assurances that at least one of the seventy-five Saudi businessmen would be present. Media press kits were prepared at no charge for the *Saudi Gazette* by Gray and Company, a public relations firm.

American firms from all over the country converged in Atlanta. This included top corporate heavyweights such as Lockheed Aircraft International; Hughes Aircraft; Citibank; Westinghouse; Boeing; Shell Oil; Aramco; Mobil; Shell; the West Company, John West's consulting outfit; American Mechanical Contractors; and the Birmingham-based multinational, Blount Inc., which by now was becoming a regular guest on the Arab trade conference circuit. Even those with little previous trade with Saudi Arabia were on hand. The sales manager of Pure Water, Inc., of Lincoln, Nebraska, was there to sell bottled water; while the head of Polycoat Systems of Hudson Falls, New York, was looking for new Saudi contracts to expand his sales of polyurethane foam insulation beyond those he was making to King Khalid Military College.[19] Several mid-level businessmen brought their preliminary contracts with Saudi Arabia, hoping to find the requisite contact, a Saudi agent, to shepherd the deal through the Saudi bureaucracy.

An outpouring of gratitude toward Young by Arab officials was clearly in evidence at the conference. "We Arabs are very loyal people. We remember things like that," commented Abdullah Alireza, the Saudi businessman who had been a guest of honor at the Birmingham conference. Michael Saba, an American who worked for OKAZ, a conference sponsor, declared, "Arabs owe a debt in eternity to Andy Young they may not be able to repay in this generation." [20]

A stellar cast of speakers awaited the businessmen at the International Tower of the Hyatt, including Secretary of Agriculture John R. Block, former Attorney General Griffin B. Bell, former Georgia Governor George Busbee, top Saudi government officials, and leading business executives representing Arab and American multinationals.

Block offered the Saudis a rather extraordinary pledge. He unilaterally promised that the United States would never invoke a food embargo against the Arabs—without eliciting any reciprocal commitment about an oil embargo against the United States. Fouad al-Farsy, Saudi deputy minister of industry and electricity, lost no opportunity to deliver a message to the "Zionist lobby." Referring to a newspaper column in the *Atlanta Constitution*, written by former Carter administration official Stuart Eizenstadt—in which he expressed his hope that the Atlanta conference would not be as politicized as the Birmingham one—al-Farsy said: "We appreciate [Eizenstadt's] advice and assume he is ready to direct it to the Zionist lobby. If they placed America's interest first, they wouldn't have to worry about the things they worry about."

But the biggest star was Jimmy Carter. In his talk "from his heart" the former President began by expressing his appreciation to two guests at the conference, John West—who, Carter said, "represented me so well"—and Andrew Young, "one of the great international leaders of all time." Carter described his visit two months before to the Middle East. He spoke of his meetings with political leaders in Egypt, Jordan, Syria, Saudi Arabia, and Israel. Yet the only political issue that he raised was the Arab-Israeli dispute. Indeed when he referred to Israel, he cited the Palestinian problem, noting their "sixteenth year under military occupation." Carter touched on no other political issue. It was as if human rights violations existed nowhere else in the Middle East. "I learned more than I have ever known as President about the suffering of the Palestinian people," Carter said as he told the audience about his trip to the West Bank and Israel, which lasted one week. Strangely enough, the man who more than anyone else had elevated the issue of human rights to the fore of the American and international political agenda, did not express any concern about the repression of Saudi Shiites, the authoritarian rule and denial of basic human rights in Saudi Arabia, the discrimination against the Copts in Egypt, or even the torture and slaughter of thousands of political and religious dissidents in Syria.

When he discussed his relationship with Saudi leaders and his visit to Saudi Arabia, Carter fawned over the Saudi hierarchy: "I'm particularly proud in a personal way of my relationship with the Saudi leaders. This has been a very gratifying thing to me both while I was President and the last two years since I left the White House." He noted that King Khalid, Crown Prince [now King] Fahd, Prince Abdullah, and Prince Sultan "have gone out of their way to be helpful in times of crisis and challenge and to provide me advice, counsel and support at times when I tried to assume some of the burdens which perhaps could have been avoided." He related how he came away so impressed from his recent visit, where he traveled 160 miles north of Riyadh into the desert to meet with the King and Saudi tribal chieftains.

Carter's praise was effusive: "I've especially come to comprehend and to appreciate the high significance of family life in the worlds of the people of Saudi Arabia and the great respect that's paid to the elderly, and also the ties that bind those of close kinship together, the extraordinary hospitality, the desire for accommodation, the need to preserve and protect the community of Islam.'' In addition, the former president said he came away appreciating the Saudi ''need for unanimity to prevent a division and separation in the Arab world.'' And he recounted his talks with Saudi leaders: ''We discussed the adverse consequences of the reductions in oil demand and the reduction in oil prices.'' (Ironically, there was probably no factor more responsible for the economic recession under Carter's reign than the rise in oil prices. Furthermore, it was the pan-Arab summits which had successfully sabotaged the expansion of Carter's greatest diplomatic achievement—the Camp David Accords.) At the end of his speech, Carter told American businessmen that if they wanted to ''form alliances that are mutually profitable in the world of Saudi Arabia, there needs to be a special effort to understand'' the Saudis and the Arab world.

Six months after the conference a wealthy businessman picked up the $50,000 tab for a benefit held on October 5, 1983, at the Sotheby Parke Bernet building in New York City to raise funds for Carter's Presidential Library in Atlanta. The businessman was the prominent Saudi sales agent, Adnan Khashoggi.

EPILOGUE

By 1984, the world oil glut, brought on by energy conservation and the development of non-OPEC oil supplies, had clearly begun to take a heavy economic toll on Saudi Arabia and the other oil producers. OPEC's share of the oil market plummeted radically from 31 million barrels per day in 1979—representing 60 percent of total world sales—to 16 million barrels per day, a 30 percent share. The price of oil had fallen below $29 a barrel by August 1984, with some countries, like Iran, providing additional discounts on the side. Saudi Arabia's 1983 oil earnings dropped to $48 billion (according to International Monetary Fund figures), resulting in a $25 billion economic deficit, forcing a slowdown in its industrialization program, and requiring a withdrawal of some of its bank deposits. Major construction contracts have been postponed, and several projects have been canceled outright by Saudi authorities. In Qatar and the United Arab Emirates, the central budgets have been sliced by as much as 40 percent.

Nevertheless, in spite of the drop in oil revenues, the flow of petrodollars to the United States is still immense. As of the end of 1983, Saudi Arabia constituted the sixth largest export market for the United States—a position it had occupied for the previous two years. Over four hundred American companies still have offices in the kingdom; and two thousand firms regularly conduct business there. The $7.8 billion in merchandise—air conditioners, cars, radios, textiles, and so forth—sold in 1983 by American exporters came from every state in the Union. Collectively, Arab purchases in 1983—$14 billion—accounted for one out of every ten dollars of total American exports. And combined with the dependence of the West in 1983 on 13 million barrels a day of Middle East oil—of which 8 million flows through the vulnerable Persian Gulf—the Arab oil producers, led by Saudi Arabia, continue to wield influence over American policy. Moreover, American dependence upon Middle East oil increased by 44 percent in the first half of 1984 (including a 68 percent increase in the use of Saudi oil); and if present trends continue, the United States will find itself in future years as precariously dependent upon Saudi oil as it did in 1973 and 1979.

Perhaps no one was more cognizant of the fact that this leverage continues to exist—in both the short and long term—than the Saudi ambassador to the United States, Prince Bandar bin Sultan. Speaking before the Washington Press Club in Washington on April 10, 1984, Prince Bandar delivered a stern warning. Unless his government was allowed to purchase the weapons it wanted, he said, Saudi Arabia and the other Arab nations would take their arms business and trade with the United States to other countries, such as the Soviet Union, France, or Britain. And those stakes are particularly high, Bandar said, pointing out that American trade with the Arab world created an additional 600,000 American jobs. The Saudi ambassador also strongly cautioned against any congressional move to transfer the American embassy from Tel Aviv to Jerusalem, Israel's capital.

The cause for Bandar's outburst was President Reagan's

decision in March to withdraw his proposal to sell Stinger anti-aircraft missiles to Saudi Arabia amid mounting congressional opposition. Portable and shoulder-fired, the Stinger missile is a terrorist's delight. According to Mark Kirk, a defense analyst for the Washington-based Center for Defense Information, the Stingers are inappropriate for Saudi oil tanker security—more suitable against helicopters and light aircraft,which explains why the Saudis are interested in using the Stingers to protect King Fahd's yacht. Expressing concern for Israel's security and a fear that the Stingers would fall into the wrong hands, Congress threatened to block the sale. The president withdrew the sale.

Yet two months later, in May, President Reagan authorized the immediate sale of the Stingers—though the number was one-third of the original 1,200 sought—by invoking a "national security" clause that has been used only once before to bypass Congress. The President claimed that Saudi Arabia needed the missiles to protect itself from Iranian air strikes. Yet according to administration sources, there was no intelligence information to indicate that the Iranians were planning to strike Saudi oil fields. The President used this extraordinary means to facilitate the sale of the Stingers simply because Saudi Arabia demanded the right to purchase them. Like the decision to approve the sale of the F-15 aircraft in 1978 and of the AWACs in 1981, the sale of the Stingers had become a "litmus test" of U.S.–Saudi relations.

At the same time, American corporations began to heed Saudi protests against the congressional resolution to move the American embassy from Tel Aviv to Jerusalem, Israel's capital. One unusual firm which lobbied against the Jerusalem move was the Encyclopaedia Britannica Educational Corporation. On April 2, 1984, a vice-president of the Chicago-based company wrote to Illinois senators Charles Percy and Alan Dixon warning that if the United States relocated the embassy, "it would have the emotional impact in the Muslim world that would result throughout the United States if some superpower were to enforce the return of Illinois to French sovereignty." Encyclopaedia Britannica Educational Corporation sells millions of dollars of educational films in all of the Arab countries of the Persian Gulf as well as in Israel.

It must be pointed out that a solid case can be made for not moving the American embassy to Jerusalem, especially since control of the city is a major issue in the Arab-Israeli conflict. Similarly, legitimate arguments can be made on behalf of selling Saudi Arabia various weapons. But the American decision-making process becomes warped when legitimate arguments turn out to be covers for other reasons—such as Saudi ultimatums or commercial self-interest. The American public, whose support is necessary for the successful implementation of any foreign policy, has a right to know who is pulling the political strings and why—especially where foreign influences are involved.

As far as the lobbying process is concerned, the American public is supposed to be protected from surreptitious manipulation by foreign interests or by Americans working on behalf of foreign interests.

The Foreign Agents Registration Act is designed explicitly to ''protect the integrity of the decision-making process'' by requiring the registration at the Department of Justice of any individual or group acting at the order, request, direction, or control of a foreign government in promoting that foreign government's political interests. According to knowledgeable lawyers and former government officials, possible violations of the Foreign Agents Registration Act have been committed by corporate executives and companies who lobbied Congress or contributed to political groups at the request of Saudi officials; and by executives and firms who lobbied Congress or contributed to political groups under the threat or inducement of losing or gaining Saudi contracts. (According to government sources, the Justice Department initiated a preliminary investigation into violations of the Foreign Agents Registration Act in March 1982, following the AWACs debate, but then suddenly dropped the inquiry.)

The breadth and scope of the petrodollar impact is beyond any legal remedy. With so many corporations, institutions, and individuals thirsting after—and receiving—oil money, petrodollar influence is ubiquitous in American society. The result is the appearance of widespread, spontaneous support for the policies of Saudi Arabia and other Arab oil producers by American institutions ranging from universities to the Congress. The proliferation of vested ties has allowed special interests to be confused with national interests.

Never before in American history has any foreign economic power been as successful as Saudi Arabia in reaching and cultivating powerful supporters all across the country. The Saudis have discovered that one quintessential American weakness, the love of money, and the petrodollar connection has become diffused throughout the United States.

NOTES

CHAPTER 1

1. These eleven are Algeria, Indonesia, Iran, Iraq, Kuwait, Libya, Nigeria, Qatar, Saudi Arabia, United Arab Emirates, and Venezuela. (Ecuador and Gabon are also OPEC members but their revenues are marginal compared to the rest.)
2. *Washington Post,* August 18, 1982.

CHAPTER 2

1. Unless otherwise indicated, all material in this chapter regarding the activities and meetings of oil company officials comes from three sources: (1) the 1974–75 hearings of the Senate Subcommittee on Multinationals of the Senate Foreign Relations Committee on "Multinational Corporations and U.S. Foreign Policy." Parts 4, 5, 6, and 7; (2) documents published by this subcommittee; and (3) interviews by this writer with oil company officials.
2. Yet, paradoxically, many Americans have not heard of the Arabian American Oil Company—Aramco. The operations of the consortium are in Dhahran, Saudi Arabia (its main offices were moved there from New York in 1952), and the four shareholders deliberately maintain its low profile. To the extent there is recognition of Aramco, many people really do not know what it is, and some even tend to think of it as an Arab

corporation. Yet, until 1980, Aramco had been a distinctly American corporation, owned and staffed by the four American oil companies. In 1980, control of the company reverted to the Saudis as they bought the assets of Aramco located in Saudi Arabia, but the company is still an American company incorporated in Delaware.

3. In 1950, Aramco's income tax payments to the U.S. Treasury amounted to $50 million. But in the following year, Aramco's income tax payments plummeted to $6 million, as a result of the oil tax credit. Since that time, the rate of taxation for the oil companies has been fantastically low. In 1972, for example, Mobil paid only 1.3 percent of its worldwide income in American taxes. And in 1976, as disclosed by columnist Jack Anderson (*Washington Post*, June 4, 1982), Aramco paid nothing in taxes on total revenues of $36 billion. The oil credit allowed Aramco to treat $24.64 billion it made in payments to the Saudis as income tax rather than royalties. And classified as an income tax, it allowed Aramco to offset dollar for dollar any U.S. income tax it might have owed. The $24.64 billion was more than enough to offset all tax liability to the U.S. Treasury.

For an excellent discussion of the history of this arrangement, see the Report by the Senate Subcommittee on Multinationals, January 2, 1975 (Part 3) and hearings, Parts 6 and 7.

4. J. B. Kelly, *Arabia, the Gulf and the West,* Basic Books, New York, 1980, p. 61.

5. Libya had actually instituted an embargo on its oil to the United States after the June 1967 war. But market conditions rendered the embargo entirely ineffectual and within months Libya had ended it.

6. Chaim Herzog, *The War of Atonement* (Jerusalem: Steimatsky's Agency Ltd., 1975), pp. 38–39.

7. Anthony Sampson, *The Seven Sisters* (New York: Viking Press, 1975), p. 266.

CHAPTER 3

1. *Time,* February 10, 1975.

2. Accounts of Malott's speech appeared in the *Lawrence Daily Journal World,* March 26 and March 27, 1975, and in the *University Daily Kansan,* March 27, 1975.

3. Declassified CIA Intelligence Report, "Problems with Growing Arab Wealth," July 1974.

Later a different type of financial crisis would emerge resulting from the overextension of commercial banks in providing loans to developing countries that would be unable to repay the debt. The specter of the domino reaction of defaults by debtor nations was actually foreseen as early as 1975 by the Senate Foreign Relations Subcommittee on Multinational

Corporations, which sought to determine the extent of major loans to foreign nations by the banks. In 1977 Karin Lissakers of the Subcommittee wrote a brilliantly researched analysis of the debt problem, which predicted exactly the type and scope of international financial problems that the world is now facing. See "International Debt, the Banks and U.S. Foreign Policy," Senate Foreign Relations Committee, 1977.

4. *Aramco World,* January-February 1977; SRI International Press Release August 7, 1975.
5. Anthony Cook, "The California-Saudi Connection," *New West,* July 3, 1978.
6. *Palo Alto Times,* August 8, 1975.
7. SRI Annual Report, 1977.
8. For an excellent profile of Bechtel, see Mark Dowie, "The Bechtel File," *Mother Jones,* September/October 1978.
9. When the petrochemical plants begin production in 1985, Saudi Arabia will emerge as one of the world's largest producers of petroleum by-products, which constitute the key ingredients for plastic, film, fertilizer, polyurethane, foam rubber, and thousands of other consumer goods and factory materials.
10. *Aramco World,* January-February 1977. Two years later, the projected gas output was halved, but the final cost may well exceed $15 billion (*New York Times,* November 17, 1979).
11. *Aramco World,* January-February 1977.
12. Ibid.
13. Ibid.
14. Tom Curtis, "Allah in the Family," *Texas Monthly,* April 1977.
15. Pharoan's purchase prompted an investigation by Congress to determine if any conflict of interest arose from the contracts that CRS had won from the U.S.–Saudi Joint Economic Commission.
16. *Wall Street Journal,* January 17, 1975.
17. *Business Week,* May 14, 1979.

CHAPTER 4

1. My sources for al-Zamel's remarks and the account of his trip are from: the *New York Times,* May 28, 1975; *Journal of Commerce,* May 28, 1975; *Chicago Sun-Times,* May 13, 1975; press release of the government of Saudi Arabia, June 1975; interviews with corporate officials who had attended al-Zamel's talks.
2. Ibid.
3. *New York Times,* May 28, 1975.
4. Letter and accompanying brochure published by the Arab Press Service

(APS). APS sponsored several talks by this economic delegation including those in Houston, New York, Los Angeles, and Chicago.

5. Arab Press Service release and transcript, June 4, 1975.
6. *San Francisco Chronicle*, April 22, 1975; *San Jose Mercury*, April 25, 1975.
7. *Times-Picayune* (New Orleans), May 20, 1975.
8. Arab Press Service, June 4, 1975.
9. Walter Henry Nelson and Terence C. F. Prittie, *The Economic War Against the Jews*, (New York: Random House, 1977), p. 65.
10. U.S. Department of Commerce letter and pamphlet, November 19, 1974.
11. Transcript of Battle's comments, *Arab Economic Review* (January-February 1975), vol. 8, no. 1, published by the U.S.–Arab Chamber of Commerce.
12. While worldwide inflation had increased food and wheat prices, their rate of increase was nowhere near as great as that of oil prices. For example, the price of food increased by 146 percent over a three-year period, 1972–74. Oil, on the other hand, rose 600 percent in just half a year as a result of OPEC's price hike. Moreover, the price of food actually declined by more than 19 percent between 1974 and 1975, while oil prices continued to rise. *OECD Economic Outlook*, July 17, 1975. Finally, the increased oil prices drained more money out of the economies of the less developed countries than did any other purchase; to a large extent the debt crisis of the Third World today is due to the actions of the OPEC cartel.
13. In an article he wrote for the *Journal of Commerce* about the same time (November 18, 1974), Battle's portrayal of King Faisal read like Saudi public relations copy: "Reluctantly, the King has accepted the fact that he and his oil are a major financial force in the world. He would have preferred that it not happen. . . . King Faisal has no political aspirations outside his borders except in a religious context."
14. Nelson and Prittie, p. 194.
15. Nelson and Prittie, p. 189.
16. *Washington Post*, January 17, 1976. The State Department even attempted to stop the Justice Department from going ahead with the suit. For an account on the State Department's and Henry Kissinger's roles, see Sol Stern, *New Republic*, March 27, 1976. Also, for a study of the early (1965–1977) corporate response to the Arab boycott, see Mark Green and Steven Solow, "The Arab Boycott of Israel: How the U.S. and Business Cooperated," *Nation*, October 17, 1981.
17. *Boston Globe*, February 19, 1975; *Newsday*, March 19, 1975.
18. *Wall Street Journal*, February 14, 1975; March 6, 1975; *Time*, February 24, 1975.
19. *New York Times*, November 16, 1975, p. 33.

20. *Kern* vs. *Dynalectron Corporation;* Brief of Appellant, Appeal from the United States District Court for the Northern District of Texas; also interview with Art Brender, attorney for plaintiff.
21. *Los Angeles Times,* November 20, 1983.
22. *New York Times,* May 11, 1975.
23. The Saudi refusal to allow blacks into Saudi Arabia apparently stemmed from fears that black Muslims would enter the country. See Nelson and Prittie, p. 79.
24. In 1975 the United States exported goods worth $1.49 billion to Saudi Arabia. In 1976 exports jumped to $2.73 billion. In 1978, a year and a half after the anti-boycott legislation was passed by Congress, American exports to Saudi Arabia zoomed to $4.37 billion. The *American Banker,* a daily newspaper for the banking world, published an article on April 19, 1978, entitled "U.S. Boycott Laws Seen Not Affecting Trade with Arabs." And a house publication of Chase Manhattan Bank, *Mideast Markets,* reported that Arab officials "take care to administer their boycott policies in such a way that they do not cast a net in which almost any American company is likely to be caught."
25. *Washington Post,* January 9, 1976.
26. *New York Times,* September 7, 23, 30, 1976; *Washington Post,* January 9, 1976.
27. *Business Week,* September 27, 1976.
28. *Washington Post,* September 27, 28, 1976; *Washington Star,* September 25, 27, 1976.
29. *Aramco World,* January-February 1977.
30. Ibid.
31. *Washington Post,* February 21, 1978.
32. *Amsterdam News* (New York), June 24, 1978; *Washington Post,* June 10, 1978.
33. *Los Angeles Times,* January 23, 1979.

CHAPTER 5

1. *Wall Street Journal,* March 29, 1982.
2. Ibid.
3. Interview with Frederick Dutton, January 21, 1982.
4. Interview with Frank Jungers, March 12, 1982.

CHAPTER 6

1. Robert Walters, "Big-Name Americans Who Work for Foreign Countries," *Parade,* June 20, 1976.
2. *Business Week,* March 14, 1975.

3. *Platt's Oilgram News*, March 14, 1975.
4. *Newsday*, April 27, 1975.
5. Address of William Fulbright before the annual conference of the Middle East Institute, October 3, 1975. Notes on Fulbright's talk provided by an observer in the audience. Fulbright's address is also quoted in *Near East Report*, October 8, 1975.
6. *New York Times*, June 12, 1974.
7. *Newsweek*, September 11, 1978.
8. *Newsweek* was not alone in omitting Fulbright's status as a registered agent of Saudi Arabia. Other periodicals also quoted Fulbright's views on Israel without mentioning that he worked for the Saudis. The most recent example occurred on May 6, 1984, when the *Washington Post* printed a lengthy profile of Fulbright, in which his critical comments about Israel figured prominently. "The Israeli lobby is the most powerful in Congress," Fulbright said. "Israel could not exist except for American aid over the years, but they may pursue policies contrary to our own interest." Nowhere in the article was Fulbright's work for the Saudis mentioned—nor the $200,000 in Saudi money that his firm took in.
9. *Columbia State*, May 16, 1978.
10. Thurmond's vote, according to South Carolina sources, may have had more to do with promises of financial help from wealthy Republican businessmen who had large textile contracts with Saudi Arabia.
11. Interview with J. Crawford Cook, February 24, 1982.
12. *Newsday*, September 5, 1978.
13. The staff member, Stephen Bryen, was allegedly overheard in a Washington restaurant in 1978 offering to provide classified documents to Israeli military officials. The accuser was Michael Saba, a former executive director of the National Association of Arab Americans. Bryen denied the accusation. The matter was dropped by the Justice Department due to lack of evidence to substantiate Saba's charge.
14. *Saudi Arabian and United States Relations: Current Prospects and Problems*, April 5, 1978.
15. Craig Karpel, "The Petro Industrial Complex," *Penthouse*, June 1978.

CHAPTER 7

1. See Hearings of a Subcommittee on Government Operations: *Federal Response to OPEC Country Investments in the United States*, Part 1, Overview, pp. 527–46.
2. *US News & World Report*, November 1, 1976.
3. Peter Grose, *Israel in the Mind of America* (New York: Knopf, 1984).
4. *Business Week*, January 23, 1978.

5. *Washington Post,* August 10, 1978.
6. *Newsweek,* June 13, 1977.
7. Steven Brill, "Connally, Coming on Tough," *New York Times Magazine,* November 18, 1979.
8. "The Intelligencer" *New York* magazine, August 25, 1980, p. 9.
9. *Washington Star,* May 29, 1976.
10. *Washington Post,* October 10, 1976.
11. Quoted in Robert Pack and Peter S. Greenberg, "In Search of Spiro Agnew," *Washingtonian Magazine,* April 1983.
12. Bankruptcy proceedings of the Atlantic International Corporation. Quoted in *Washington Post,* April 8, 1978.

CHAPTER 8

1. *Petroleum Intelligence Weekly,* July 17, 1977.
2. *New York Times,* December 18, 1976.
3. Multinational Petroleum Companies and Foreign Policy, Hearings before the Subcommittee on Multinational Corporations of the Committee on Foreign Relations. United States Senate, 93rd Congress, second session, June 20, 1974.
4. *Ibid.*
5. *New York Times,* June 23, 1977.
6. *New York Times,* May 12, 1977.
7. "OPEC: History and Prospects of an Oligopoly," Report No.n286, January 21, 1976. Bureau of Intelligence and Research, U.S. Department of State. Released by the chairman of the House Subcommittee on Commerce, Consumer and Monetary Affairs, Congressman Benjamin S. Rosenthal, June 23, 1980.
8. *Washington Post,* January 24, 1978.
9. Letter quoted in follow-up article by Hersh that appeared in the *New York Times* on February 8, 1978.
10. If there was any legal substance to the charge that company transfer or committee publication of oil field data was an invasion of Saudi privacy, the State Department's legal adviser's office would surely have gotten involved. But it never did.
11. *The Future of Saudi Arabian Oil Production,* staff report to the Subcommittee on International Economic Policy of the Senate Foreign Relations Committee, April 1979.
12. The subcommittee's investigation was not the only federal investigation to experience Saudi intervention. In December 1983, Assistant Attorney General William Baxter, a Reagan appointee, dropped a six-year antitrust investigation of Aramco and the international oil market. The investiga-

tion had been started because of the critical and alleged anti-competitive role played by the oil companies in restricting oil production to set the price. (Because the oil exporting nations function primarily as oil producers, they depend upon the companies to purchase, refine, market, and distribute their oil. The net result is that the companies are the ones that adjust the level of production among OPEC and non-OPEC members; in effect, the companies serve to maintain the balance between OPEC supply and free-world demand. Within this framework the arrangement between Saudi Arabia, the largest oil exporter, and the Aramco consortium is chiefly responsible for forestalling the emergence of competition and the collapse of cartel pricing.)

When the case was terminated in December 1983, the Justice Department claimed they could find no evidence demonstrating that "under current market conditions, the Aramco partners could exercise market power." Yet, after 1977, the Justice Department was never able to obtain key documents from the Aramco partners and acknowledged that they were forced to rely on "press reports" to conclude "that the Aramco partners had neither the power nor the incentive" to control Saudi output. Without key documents, like the 1977 Aramco takeover agreement, there was no way in which the Justice Department could have made such a determination.

What prevented the Justice Department from securing these vital pieces of evidence? The Saudi Arabian government warned the Carter and Reagan administrations that unless the potentially embarrassing investigation was terminated, Saudi Arabia would retaliate against the United States. On a visit to Riyadh in December 1979, Treasury Secretary William Miller was told unambiguously by oil minister Yamani that he expected Treasury's assistance in quashing the investigation. Treasury immediately relayed these "concerns" to the Justice Department.

Treasury was not alone in getting involved. Fearing Saudi ill will, the National Security Council and the State Department actively sought to stop Justice from pursuing its subpoenas. According to government sources and documents obtained under the Freedom of Information Act, five separate bureaus within the State Department became simultaneously involved in ensuring that the Justice Department was apprised of the foreign policy "sensitivity" of its investigation.

In March 1980, Deputy Attorney General John Shenefield, who was in charge of the investigation, traveled to Saudi Arabia to obtain Saudi permission for the companies to release certain documents pertaining to the investigation. All along, the companies had argued that Saudi Arabia would not allow them to comply with the Justice Department's subpoenas—despite the fact that the documents were wholly owned and possessed by the companies. But Yamani would not budge, as Shenefield left Saudi Arabia empty-handed and the investigation was put on the back burner.

When the Reagan administration assumed office, the new Justice Department appointees began to review the investigation. But the vital oil company documents were still missing. There was no further effort made to secure these documents. Saudi officials informed the Reagan administration that this investigation was a "thorn" in U.S.–Saudi relations. And so, rather than incur Saudi wrath, the Justice Department dropped the subpoenas. (See Edwin Rothschild and Steven Emerson, "Born Again Cartel," *New Republic,* November 5, 1984.)

CHAPTER 9

1. Interview with David Fanning, May 26, 1984.
2. Among the numerous British industrial leaders who condemned the broadcast of the film was Max George, sales director of the British Hovercraft Corporation, Saudi Arabia's sole supplier of Hovercraft. As reported by the *Guardian* on April 11, 1980, George said: "ATV did not use much foresight or responsibility about the possible effects their production could have. We feel that most British firms likely to be affected by an export ban must feel extreme displeasure towards ATV." Moreover, in an indication of the extent of Saudi influence in British financial circles was the *London Sunday Times* revelation on April 13 that the person who interceded with ATV on behalf of the Saudi royal family was banker Philip dc Zulucta, chairman of the merchant bankers Antony Gibb and private secretary to three successive Tory prime ministers: Eden, Macmillan, and Douglas-Home. Moreover, Sir Philip was also the brother-in-law of Lord Windlesham, the managing director of Britain's ATV, a co-producer of the film.
3. *New York Times,* April 11, 1980.
4. *8 Days,* May 30, 1980.
5. *Daily Mail,* April 14, 1890.
6. *Al-Jazirah,* quoted in the *Washington Post,* April 24, 1980.
7. *New York Times,* April 24, 1980.
8. Associated Press, April 23, 1980; *Washington Star,* April 24, 1980.
9. Associated Press, May 22, 1980. *Washington Star,* May 22, 1980. Parsons officials would not confirm this ban.
10. Yet, in the six months to a year following the broadcast, an indeterminable number of contracts were believed to have been canceled or postponed according to British trade officials in interviews in 1984.
11. In a bizarre twist, Michael Jay Solomon, head of Telepictures, which owned partial worldwide rights to the film, apparently considered exploiting the Saudi anger by offering to sell them the rights to the film in exchange for as much as $10 million. The Saudis would then have been able to suppress the broadcast. No such deal ever materialized. When *Death of a Princess* producer David Fanning met with Solomon at a European film

festival soon thereafter, Fanning exploded at Solomon, warning him that if he ever tried to sell the rights, Fanning would try to ensure that he never again participated in a PBS project.

12. *Washington Post,* April 24, 1980; *New York Times,* April 25, 1980.
13. Associated Press, April 24, 1980.
14. Associated Press, April 8, 1980.
15. Hearings before the Senate Foreign Relations Subcommittee on Multinationals, Part 5, pp. 160–61.
16. Associated Press, April 28, 1980; *Wall Street Journal,* May 12, 1980.
17. "All Things Considered." Interview by Sanford Ungar on National Public Radio, May 8, 1980.
18. Telephone Interview with Admiral Thomas H. Moorer.
19. *Washington Post,* May 7, 1980.
20. The "list of particulars," whose existence had never before been revealed, was disclosed to me by producer David Fanning. He said that he actually felt relieved and complimented after he saw the complaints because they were so trivial. For example, one item complained that the license plates used in the film were Egyptian and not Saudi.
21. *Daily News* (New York), May 9, 1980.
22. *Grand Rapids Press,* May 11, 1980.
23. PBS records.
24. Transcript of editorial, WJBL, May 12, 1980.
25. *Minneapolis Star,* May 12, 1980.
26. Interview with Fred Rebman, January 12, 1981. In addition, columnist Dave Montoro of the *Florida Times-Union* reported at the time that in deciding not to broadcast the film, Rebman believed it was "time when you say, 'These are our allies,' " *Florida Times-Union,* May 13, 1980.
27. *Columbia Record,* April 25, 1980.
28. Associated Press, May 6, 1980.
29. *Columbia State,* May 10, 1980.
30. *Columbia Record,* February 23, 1978.
31. *Birmingham News,* May 12, 1980.
32. *Birmingham News,* May 12, 1980. The report of internal deliberations of AETC is based on depositions of AETC officials in *Muir* vs. *Alabama Educational Television Commission,* 656 F.2d 1012, N.D. Ala. 1980.
33. PBS records.
34. *Barnstone* vs. *University of Houston,* 514 F. Supp. 670 (S.D. Tex. 1980). Court depositions and records.
35. It is interesting to note that not all of Saudi Arabia's wrath was reserved for the West. According to Daniel Pipes, after "the Syrians antagonized Saudi leaders by having the state-controlled press declare that the film 'Death

of a Princess' was less anti-Islam than the Saudis claimed, Riyadh withdrew its ambassador from Damascus in protest . . .'' Daniel Pipes, *In the Path of God: Islam and Political Power,* Basic Books, New York, 1983.

CHAPTER 10

1. Letter to Senator John Sparkman, chairman of the Senate Foreign Relations Committee from Secretary of Defense Harold Brown, May 9, 1978; State Department comments on the AIPAC memorandum in response to a request of Congressman Lee H. Hamilton, February 16, 1978; letter to Congressman Gerry E. Studds from Under Secretary Lucy Wilson Benson for Security Assistance, Science and Technology, March 21, 1978.
2. Ibid.
3. *New York Times,* June 17, 1980.
4. As a member of the staff of the committee working closely with its chairman, Frank Church, I made over a dozen inquiries, by letter and by telephone, to the White House and State Department regarding the newspaper reports. Without exception, all of the responses contained vague denials of any intention to sell new equipment to the Saudis.
5. "Saudis Claim U.S. Pledge of Arms Sale," *Washington Star,* July 18, 1980.
6. Insofar as I had responsibility for writing the first draft, and all subsequent changes were made at the staff level of the Senate Foreign Relations Committee, it appears that some of my colleagues were secretly doing the bidding of the White House at the behest of the Saudis. This was told to me by a member of the Carter White House who said that his administration intended to resubmit the sale immediately following the November elections. The new language, he told me, was designed to allow the senators to reverse their positions without being accused of flipflopping.
7. The legislative veto was ruled unconstitutional by the U.S. Supreme Court in 1983. The status of Congress's right to disapprove arms sales is presently unclear.
8. *Time,* October 5, 1981.
9. *New York Times,* October 27, 1981.
10. *Washington Post,* October 29, 1981.

CHAPTER 11

1. Strangely enough, I had met Dabbagh in 1980 in my capacity as a staff member of the Senate Foreign Relations Committee. Accompanied by two representatives of the multinational Dravo Corporation, who had set up

the lunch, I met Dabbagh at the Fairfax Hotel. In the first several minutes Dabbagh told me how much power I wielded as an adviser to a senator and then launched into a passionate argument as to why Israel was the aggressor state in the Arab-Israeli dispute. But in a casual aside Dabbagh revealed that as a result of the recent resignation of U.N. Ambassador Andrew Young over his contact with the PLO and the much-celebrated visits to the PLO and the West Bank by the Reverend Jesse Jackson and District of Columbia Delegate Walter Fauntroy, Saudi Arabia intended to play an active behind-the-scenes role in the 1980 elections by providing funding to black groups.

2. *Wall Street Journal,* October 22, 1981, p. 1.

3. Press coverage of the AWACs debate concentrated on Jewish lobbying efforts while little, if any, attention focused on corporations. The principal exception to this record were stories by reporter Paul Taylor of the *Washington Post,* October 31, 1981; columnist Hobart Rowen of the *Washington Post,* November 8, 1981; syndicated columnist Jack Anderson, October 26, 1981; and reporter C. David Kotok of the *Omaha World Herald,* October 29, 1981.

4. *Wall Street Journal,* October 22, 1981.

5. Other Saudis also played a pivotal role in directing efforts, meeting regularly with White House, State Department, and Defense Department legislative affairs officials. In one meeting that took place in Riyadh in early September, Saudi Defense Minister Sultan complained bitterly to American Ambassador Richard Murphy about public statements made by certain Reagan administration officials. According to a Pentagon document obtained under the Freedom of Information Act, Sultan bitterly complained that "recent statements in U.S. media" made by Reagan officials were "harmful to our common purpose," and demanded that they be stopped.

6. *Mideast Markets,* September 4, 1981.

7. *Washington Post,* September 28, 1981.

8. *Time,* October 12, 1981.

9. An examination of American, Middle Eastern, and Saudi trade publications which cover business developments in the region and monitor major contracts awarded by Saudi Arabia confirms this twenty-three-day contract suspension.

10. *Newsweek,* August 24, 1981.

11. Advertisement by the Brunswick Corporation in the *New York Times,* February 2, 1982. Quote about Alibrandi is from a *Washington Post* article on a Securities and Exchange Commission probe of the Whittaker Corporation's ties to Saudi Arabia, February 24, 1980.

In February 1983, Whittaker's contract in Saudi Arabia was increased to $1.4 billion. Whittaker's spectacular success in obtaining Saudi hospital contracts was achieved partially through the well-connected sales agents, such as Adnan Khashoggi and members of the royal family. Suddenly, in December 1983, Whittaker was informed by the Saudi government that its hospital contracts would be terminated. The Saudi decision appeared to reflect a new policy designed to open up contract renewals to competitive bidding. In the end, Alibrandi's pro-Saudi activities did not save his contracts. The Saudi decision not to renew the contracts resulted in Whittaker's stock plummeting in value.

12. Interview with Dorothy Lorant, vice-president for public relations and advertising for the Greyhound Corporation. In a subsequent letter to me, Lorant wrote, "we at Greyhound feel that writing to one's Congressmen is a forum that should be utilized by corporations as much as by private citizens or by entities as diverse as the AFL-CIO or the Sierra Club. . . . I want to make it absolutely clear to you that as far as this corporation is concerned our action in writing was neither at the suggestion of any business contacts and neither was it with a view to obtaining privilege down the road."

13. Unless otherwise noted, all quotations are taken from the interviews I conducted for an article I wrote for the *New Republic,* "The American House of Saud: The Petrodollar Connection," February 17, 1982.

14. *Washington Post,* May 23, 1983.

15. Notwithstanding Shultz's statement, Bechtel, according to a Saudi-connected consultant, was told by key Saudi officials in 1981 that they expected the company to be aggressive in its lobbying campaign. In 1978, this source said, Bechtel's passive role during the debate over the sale of F-15 aircraft to Saudi Arabia had irritated the Saudis. During the AWACs campaign, Bechtel was placed on notice that its efforts would be watched.

16. Letter to the *New Republic,* February 29, 1982, in response to my article, "The American House of Saud," February 17, 1982.

17. Memo from Baker, June 4, 1982, to National Affairs Committee of the Association of Wall & Ceiling Industries. The memo enclosed a letter written by Baker on June 2 to Wayne Vallis, Office of Public Liaison in the White House.

18. The other members of the consortium were Southern Union Refining Company (Texas); Pester Refining Company (Kansas); Marion Corporation (Alabama); and Rock Island Refining Corporation (Indiana).

19. *Political/Finance Lobbying Reporter,* November 1982.

20. *Time,* November 7, 1981. *Time* never reported on the extraordinary corporate telex, despite the fact that the trip was reported in a special column by its publisher. In fact, the only mention in *Time* of any specific corporation that lobbied for the AWACs sale during the months of September and October was a reference in the November 9, 1981 issue to a "pro-

AWACs call from Mobil Oil President William Tavoulareas to Democrat David Pryor of Arkansas.''

21. *Harvard Crimson,* February 18, 1982.
22. Letter from Matina S. Horner to Dr. Robert Kravetz and Mrs. Sherry Leonard, the Jewish Federation of Greater Phoenix. March 23, 1982.
23. Letter from Thomas J. Watson, Jr. to Dr. Robert Kravetz and Mrs. Sherry Leonard, the Jewish Federation of Greater Phoenix. March 22, 1982.
24. *Washington Post,* October 29, 1981.
25. *Omaha World Herald,* October 29, 1981.
26. *Wall Street Journal,* October 29, 1981.
27. For a discussion of the applicability of the Foreign Agents Registration Act, see Epilogue.
28. *London Times,* February 28, 1982.

CHAPTER 12

1. Sources for this chapter included several high-level government officials. They had unique access to the inside workings of the State Department. In addition, I acquired information on documents from State Department materials obtained through a Freedom of Information Act request and from documents provided to me by a government source.
2. In fact, the only suggested changes came from one of Long's immediate superiors in the Bureau of Intelligence and Research, who recommended that Long make five deletions; this involved taking out a few names and several half-paragraphs. His two considerations in making these deletions were: any possible harm to U.S.–Saudi relations and any potential impediment to the flow of intelligence material.

   In addition to the selective de facto declassification of the secret study, Long may also have violated State Department regulations governing outside publishing by its employees. Section C of Foreign Affairs Manual regulation 628.2 stipulates that clearance of outside writings will not be granted until "all material . . . which is inaccurate" has been deleted.
3. It is interesting to note that in one place an insertion in the CSIS version may have stemmed from an attempt to please a particular Saudi citizen. In the classified report two paragraphs focused on the rise of young, prominent Saudis, some of whom had been educated in Europe and the United States. The names of four such prominent Saudis—the director of operations of the Air Force and three deputy ministers—were mentioned as examples. In the corresponding CSIS paragraph, the sample was enlarged to six Saudis. Who were the two new names who were fortunate enough to be added to this prestigious group? The director general and deputy director of the Royal Commission for Jubail and Yanbu. And the

Royal Commission happens to be the former employer of the Booz Allen & Hamilton consulting operation in Saudi Arabia, which is where Shaw worked after leaving the State Department in 1977.

4. The prominent think tank, even though it was not aware of the deception, was able to take credit for generating a major study on a hot topic. But in light of the multimillion-dollar contracts in Saudi Arabia held by firms such as Avco and Booz Allen, an ancillary issue is whether CSIS should have solicited or accepted money for research on the Middle East from firms that have such an overriding interest in generating pro-Saudi material.

In the next few years, however, CSIS officials made a direct pitch for Saudi money. According to sources, CSIS officials traveled to the Middle East in early 1984 where they met with a leading member of the Saudi royal family. A very large donation was sought from the prince; a major study on the Middle East was offered in return by CSIS. By September 1984 CSIS had been promised $100,000 from this donor.

In addition, CSIS officials circulated a highly unusual memorandum to various staff members in August 1984. The memorandum—written on blank stationery— solicited a "specialist" on the Industrial Revolution to assist a young Saudi prince working on his Master's program. In exchange for "imparting information," "editing his papers," and "spending up to forty-five days in Saudi Arabia," the CSIS memo noted that the "Royal Family will be happy to pay all expenses—air fares, room and board—and will probably contribute a stipend as well."

5. State Department sensitivity to the opinion of Arab oil-producing countries has sometimes bordered on the absurd. On October 20, 1980, almost three weeks before the national elections, the Senate Foreign Relations Committee received a special memo from the State Department that had been transmitted from the American chargé d'affaires in the American embassy in Abu Dhabi. The U.S. official suggested that his memo be brought to the attention of Frank Church, the chairman of the committee. And what was in this memo? A translation of a letter written to the American ambassador in Abu Dhabi by Dr. Mohammed Mehdi in which Mehdi charged that unspecified public "statements" made by Church constituted "an extreme provocation for the peace-loving states" and a "clear violation of all the international laws." The chargé d'affaires described Mehdi as the head of a respectable organization "established to help bring the American and Arab peoples closer together."

It was clear from the chargé d'affaires's memo and the accompanying State Department note that some type of apology or clarification to Mehdi was expected of Church. Church, however, already had contact with Mehdi before receiving the State Department memo. In March and April 1980

Mehdi had published a half-page "open letter" to Church in three Idaho newspapers. The letter accused Church—who at the time was involved in a tough and bitter reelection campaign in which his chairmanship of the Foreign Relations Committee did not play well to Idaho's conservative voters—of succumbing to "Zionist pressure" and of supporting $5.2 billion in aid to Israel. Calling for Church to suspend aid to the "Zionist State" for one week, Mehdi wrote, "If this amount were to be cut off to Israel and spent in Idaho, Israel will stop taking more land and Idaho's economic ills and high rate of unemployment might be eliminated." Mehdi also flew out to Idaho and challenged Church to a public debate. Mehdi's activities received much publicity in the Idaho media—and placed Church on the defensive.

Mehdi's interference in American political campaigns was not, unfortunately, new. He had also taken out ads in 1980 against Senators Bob Packwood of Oregon and Henry Jackson of Washington. As far back as 1964, Mehdi had taken out ads in New York newspapers attacking Senate candidates Kenneth Keating and Robert F. Kennedy. In 1964, deportation proceedings had been initiated against Mehdi—an Iraqi citizen—by the Immigration and Naturalization Service for "engaging in activities contrary to the best interests of the United States." Mehdi, who has admitted getting funds from Arab governments, subsequently gained American citizenship. After Robert F. Kennedy was shot, Mehdi wrote a book in which he justified the assassination.

CHAPTER 13

1. *Washington Post,* September 1, 1981.
2. Associated Press, October 25, 1981.
3. Talcott W. Seelye, "Can the PLO Be Brought to the Negotiating Table?" *American Arab Affairs,* Summer 1982, no. 1. The article was based on a speech Seelye delivered to the Tenth Annual Convention of the National Association of Arab Americans.

   That Seelye would characterize the PLO as having rejected terrorism on the basis of these examples was rather curious. Some American intelligence experts say that PLO members were involved in the killing of Ambassador Meloy and that the PLO actively assisted the Iranian "students" who had taken over the American embassy. Finally, even if Seelye had not been apprised of such information, surely he had observed that the PLO had continued, and continues, to take credit for and openly applaud murderous acts of terrorism against Israeli and European (mostly Jewish) civilians.
4. Seelye's comments and the account of the Amherst-sponsored debate are

based on an interview with *Amherst Student* reporter Andrew J. Delaney, who covered the debate for the student newspaper, and on interviews with other persons who attended the debate. Some of the quotations appeared in an article Delaney wrote for the *Amherst Student,* October 21, 1982.

5. In an interview with me on July 23, 1984, at the Cosmos Club in Washington, I asked Seelye about this comment. He said, "I never said Jews should register as foreign agents." When I asked Seelye whether he said that "certain Jewish groups should register as foreign agents," he said to me, "I didn't say anything about registering." However, in a letter published by the *Near East Report* (December 17, 1982), which Seelye had sent to correct an earlier article which he felt contained inaccuracies, Seelye admitted that he had said that the government should "require certain Jewish groups lobbying for Israel to register as foreign agents." In addition, two other persons attending the Amherst Alumni debate recall Seelye making this comment.

6. *Washington Post,* July 17, 1983.

7. *Christian Science Monitor,* May 18, 1982. In the article, Seelye declared, among other assertions, that "[I]t is an historical fact that Syrian guns only started firing down upon Israelis after Israel violated the 1948 armistice agreement by moving into the demilitarized zone created by this agreement and located at the base of the Golan Heights." Perhaps to Seelye, this is an historical fact; to others, it was the Syrians who first shelled into Israeli territory. Besides, Seelye neglects the overriding fact that it was Syria that attacked the new Jewish state after the Israeli declaration of independence in 1948.

8. Transcripts of "The MacNeil-Lehrer Report," June 7, 1982; May 6, 1983; July 19, 1983; January 3, 1984.

9. Information about the trips comes from internal Advest materials and interviews with Advest clients. Incidentally, though Israel is often one of the stops, the real attraction lies in the visits to the Arab countries, of which Saudi Arabia has been the most popular, as well as the most frequently visited. In fact, the only reason for Israel's inclusion, according to an investment banker who is an Advest client, is to allow Advest to maintain its political balance.

10. This section is based on news accounts in the *International Herald Tribune,* October 5, 1981, and the *Oil Daily,* September 29, September 30, 1981; copies of speeches, transcripts, and other internal materials provided by the *Oil Daily;* and an interview with Joseph Fitchett, a reporter for the *International Herald Tribune,* who covered the conference.

11. In his prepared comments, besides dwelling extensively on how politically costly Saudi Arabia oil policies were, Akins focused much of his ire on the "omnipotence of the [Israel] lobby," the conspiratorial campaign

orchestrated by the "enemies of Saudi Arabia" and "racist" members of the press. He specifically accused "Jewish writers" Joseph Kraft and William Safire "who are quick to condemn any criticism of Israel as racism or Nazism" of being "quite prone to group all Arabs as the 'enemy.' "

Akins also threw in Herb Block, editorial cartoonist of the *Washington Post,* into the category of "enemies of the Arabs," comparing his cartoons and Kraft's and Safire's writings as similar to that of one of Adolf Hitler's top aides: "It is doubtful if any of these writers or cartoonists recognizes himself as a racist, it is not an epithet one applies to oneself. If one hates the Arabs, he thinks it is because they are worthy of hate; if one condemns the Arabs, it is because they are inferior. This, in the view of the enemies of Arabs, is not racism; it is merely recognition of fact. Dr. Goebbels's definition of his views of the Jews did not vary significantly from this."

What especially outraged Akins was that anti-Arab sentiment was "not only tolerated but encouraged," particularly by Jewish writers. He asserted that "a vicious anti-Jewish cartoon, an overtly racist anti-Jewish article, would bring forth an immediate lawsuit which the defendant would surely lose"—in an analogy that subtly invoked a stereotypical image of the Jewish lawyer. Akins's written text was distributed to the conference participants.

12. Even before he became ambassador, Akins—as a high-ranking State Department official—delivered a speech in June 1972 before an OPEC meeting that some observers credit with encouraging oil producers to increase their prices. Akins, in an impromptu address—something of a hallmark for him—told the gathering that oil prices were "expected to go up shortly due to the lack of short-term alternatives to Arab oil." His prediction, according to a Canadian representative at the meeting, A. P. H. Van Meurs, was "tantamount" to advocating that the Arabs raise the price of oil to $5 a barrel—advice that was readily followed. *New York Times,* March 20, 1983.

13. Aramco document printed in U.S. Senate Subcommittee on Multinationals Hearings, "Multinational Corporations and United States Foreign Policy, Part 7, February 20, 21; March 27, 28, 1974, p. 517.

14. *Saudi Report,* May 25, 1981.

15. The conference was sponsored by the American Arab Affairs Council. For the full story of this conference, see Chapter 18.

16. Reprinted in *Bergen County Record,* March 26, 1979.

17. *New York Times,* September 13, 1983.

18. This section is based on interviews with Adam Cox, a reporter for the *Indiana Daily Student;* another student who attended Killgore's lecture; and a tape of Killgore's talk.

19. "Call for a Founding Convention for the Creation of a Holy Land State," a two-page flyer distributed in November 1982, signed by Haviv Scheiber, "Chairman, Holy Land State Committee"; Rabbi Elmer Berger, "Director, Jewish Alternatives to Zionism"; Dr. John Davis, "Former Commissioner, UNRWA (United Nations)"; Andrew Killgore, "former U.S. Ambassador to Qatar"; Mark Lane, "Director, National Council on the Middle East"; and Grace Halsell, "author and journalist."
20. *Saudi Report,* September 6, 1982. Description of Holy Land State Committee's goals quoted from the committee's publications.
21. *Spotlight,* August 23, 1982. *Spotlight* is published by the Liberty Lobby.
22. Details on Killgore's American Overseas Political Action Committee are on public file at the Federal Election Commission.
23. Even if the media allow themselves to be used by these former ambassadors, they have the responsibility to inform their readers of the commercial connections of the commentators. The public has a right to decide for itself whether the views of Talcott Seelye, defending Syrian policies and attacking Israel, are disinterested analysis or might be linked to the trips he takes to see President Assad in behalf of investment bankers.
24. *Parade,* November 28, 1982.

CHAPTER 14

1. *Washington Post,* May 17, 1983; January 9, 1982.
2. Other oil company officials denied altogether that they had engaged in political activities at Saudi Arabia's behest. Herbert Schmertz, vice-president of Mobil, in a letter to the Senate Foreign Relations Committee, denied that Mobil had initiated its pro-Saudi ads under pressure from the Saudis. The June 1973 Mobil pro-Saudi advertisement in the *New York Times,* Schmertz wrote, "was clearly directed to serve the legitimate interests of the United States and was not directed against the interests of any people or country, nor was it designed to serve Mobil's interests. We spoke up when we did and as we did because we felt someone had to."
3. In a subsequent letter dated May 19, 1982, to the *New Republic,* where revelation of the Aramco contributions first appeared, Knight disputed my characterization of the Aramco funding campaign: "Mr. Emerson contorts Aramco's motives in contributing very modest sums to a limited number of American educational and cultural institutions offering courses on the history and civilizations of the Middle East into an attempt to 'manipulate' American public opinion." In addition, the four oil companies, whose officials were involved in devising the program and all of whom had to approve the donations, refused to comment.
4. Interview with Frank Jungers, March 12, 1982.
5. Interview with State Department official conducted for the *New Republic,*

"The Aramco Pipeline," May 19, 1982. All other unattributed quotes in this chapter are based on interviews conducted for this article.

6. *Washington Post,* January 9, 1975.
7. According to ANERA publications and an interview with an ANERA source.
8. *Washington Post,* January 9, 1975.
9. Milton Viorst, "Building an Arab American Lobby," *Washington Post Magazine,* September 14, 1980.
10. *Wall Street Journal,* September 7, 1978; April 10, 1984.
11. *Boston Phoenix,* January 3, 1984. Cockburn, known for his biting criticism of the establishment press and his criticism of Israel, was suspended from the *Voice* for not revealing the grant. He soon left to join the *Nation.*
12. *Washington Post,* November 29, 1982.
13. *New York Times,* December 9, 1982.
14. *Boston Phoenix,* March 1984.
15. *Washington Post,* January 9, 1975.
16. When I wrote my initial article in the *New Republic,* I interviewed several AMEU officials about sources of funding. Executive Director John F. Mahoney said he did not want to release this information as he considered it confidential. He also said that Aramco has contributed "substantial sums, but there are no strings attached. All we do is send them accountability reports." AMEU's vice-president, Henry G. Fischer, who is also curator in Egyptology, Metropolitan Museum of Art, said that Aramco has provided money but that "it wasn't enough." In addition, he said, "AMEU does not take any foreign money." AMEU President Jack B. Sunderland, of American Independent Oil, said, "I am sure that Aramco's contribution is less than 50 percent. And if any foreign contributions were given, they were minor gifts—amounting to almost nothing."
17. The pro-Arab bias in SAIS can be seen from an internal document, dated September 18, 1978, which listed various criteria in the university's search for the field director of the Middle East program. One of the criteria stated that the "ideal candidate . . . should have a comprehensive grasp of Third World perspectives on international questions generally, and of the particular Arab view of these perspectives, and be able to articulate those perspectives sympathetically, if not necessarily approvingly, to students and faculty."
18. Interview with Frank Jungers, March 12, 1982.

CHAPTER 15

1. One of AET's "white papers" focuses on "Zionist mythology." This booklet purports to disprove the "Zionist claim for a lengthy, unbroken

bond between Jews and the land of Palestine'' and to ''illustrate the beginning of a process still continuing of systematic expulsion of one people from their country by another people who covet their land.'' The booklet also predicts that Israel will expel all ''non-Jews.'' This ''white paper'' was written by the chairman of AET, Edward Henderson, a former British ambassador to Qatar.

2. At the conclusion of one especially vitriolic conference I attended in Birmingham, I spoke to a Palestinian economist who had been one of the speakers. He said to me, ''This kind of talk [about Jewish conspiracies] is dangerous. You know, it hurts my cause as well.''

3. St. Louis firms have been particularly active in exporting to Saudi Arabia. Over thirty companies have substantial contracts with the kingdom, including Banquet Foods Co. (frozen pies); Emerson Electric (electrical appliances and fixtures); Lincoln St. Louis (gasoline station equipment); Pet Industries (Mexican and other prepared foods); and the Seven-Up Company.

4. Duke University memorandum, ''Description of Islamic and Arabian Studies, Duke University,'' December 15, 1977.

5. Anthony Cook, ''The California-Saudi Connection,'' *New West,* July 13, 1978; *Aramco World,* May-June 1979.

6. Anthony Cook, ibid.

7. Internal USC document: ''Agreement for the Establishment of the Middle East Center at the University of Southern California,'' September 14, 1978.

8. *New York Times,* August 28, 1979; L. J. Davis, ''Consorting with Arabs,'' *Harper's,* September 1980.

9. *New York Times,* August 28, 1979.

10. *New York Times,* July 17, 1979.

11. *New York Times,* August 28, 1979.

12. Duke memorandum, op. cit.

13. Associated Press, *Durham Morning Herald,* May 31, 1982.

14. Annual Report, Islamic and Arabian Studies Program, 1977–1980; *Middle East Economic Digest,* October 5, 1979.

15. *Duke Chronicle,* October 1, 1979.

16. Duke memorandum, op. cit.

17. Telephone interview with Dunn, February 23, 1982.

18. *Hoya,* August 30, 1975; September 10, 1976; December 10, 1976. *Washington Post,* October 10, 1980. Annual reports of the Center for Contemporary Arab Studies.

19. *Washington Post,* November 5, 1977.

20. *Washington Post,* February 24, 1981.

21. *Hoya,* August 26, 1978. After returning the $50,000 to Iraq, Healy wrote to Iraqi officials: ''I hope . . . we can continue our conversations and

that it will be possible for the University to return to the generosity of the Iraqi government in the future and ask for a gift for which full credit can be given to the government which gave it.''

22. *Hoya*, February 1, 1980.

23. One lecture in the 1982–83 academic year was entitled "Zionism: From the Deir Yassin Massacre to the Sabra and Chatilla Massacres.''

24. "A Proposal to Encourage Programs of Arab Studies in American Private Colleges,'' confidential document, February 1977.

25. The negotiations between the State University of New York at Stony Brook and the Saudi government were revealed by Sol Stern in "A SUNY Campus Looks for Petrodollars: Saudi Brook,'' *Village Voice*, July 10, 1984.

26. *New York Times*, January 28, 1984.

27. Smithsonian officials implied that the artifacts from the Rockefeller Museum belonged to the Jordanians, who controlled East Jersualem between 1948 and 1967. In fact, the museum was built in the 1920s and operated independently in British-controlled Palestine. During the 1948 Arab-Israeli War, Jordan captured East Jerusalem and nationalized the museum.

28. In early 1982, the Metropolitan abruptly reversed its earlier decision to sponsor the Israeli exhibition. In the public outcry that followed, however, the Metropolitan again reversed its position and agreed to host the exhibition. But at that point, the Smithsonian had already entered the picture and announced it would hold the show. When the Smithsonian announced its cancellation in January 1984, the Metropolitan volunteered to host the exhibition.

CHAPTER 16

1. The office of International Bank and Portfolio Investment of the Department of the Treasury.

2. Figures obtained from unpublished Department of Treasury report prepared for the Subcommittee on Commerce, Consumer, and Monetary Affairs of the House Committee on Government Operations, April 1983.

3. The executive branch's policy of keeping secret such investment has been dubbed the "$100 billion understanding" by writer Tad Szulc in an excellent article about petrodollars, "Recycling Petrodollars: The $100 Billion Understanding,'' *New York Times Magazine*, September 20, 1981.

4. "Changes Needed to Improve Government's Knowledge of OPEC Financial Influence in the United States,'' Report by the Comptroller General to the Congress, General Accounting Office, December 19, 1979, p. 24.

5. Draft memo by Bureau of Economic Analysis of the Department of Commerce, November 28, 1978, published in "The Operations of Federal Agencies in Monitoring, Reporting On, and Analyzing Foreign Investments in the United States,'' Hearings of the Subcommittee on Com-

merce, Consumer, and Monetary Affairs of the House Committee on Government Operations, 1979, Part 5, Appendices, p. 239. (Hereafter called Hearings of Subcommittee on Commerce, Consumer, and Monetary Affairs, 1979, Part 5).

6. "The Adequacy of the Federal Response to Foreign Investment in the United States," Twentieth Report by the Committee on Government Operations, August 1, 1980. The restrictions on the use of country-by-country OPEC data by the board of governors was revealed by an internal Federal Reserve memo from T. J. Giletti, Balance of Payments Division, to Mr. Meek. Subject: Saudi portfolio investments in the United States, March 9, 1977.

7. Treasury memorandum, from Deputy Secretaries Junz and Widman to Under Secretary Solomon. Subject: Iran consultative group, published in Hearings of the Subcommittee on Commerce, Consumer, and Monetary Affairs, 1979, Part 5, p. 238.

8. Hearings of the Subcommittee on Commerce, Consumer, and Monetary Affairs, 1979, Part 2, p. 80. Particularly shocking was the fact that Treasury has never been able to find any records of any written Treasury communication, conversation, or agreement between Treasury and the Saudis regarding Simon's agreement with the Saudis on this subject.

9. Unclassified summary of cable from U.S. embassy to Department of State, June 1979, published in "Federal Response to OPEC Country Investments in the United States," Hearings of the Subcommittee of Commerce, Consumer, and Monetary Affairs, September 1981, Part 1, p. 552. The name of the Middle East country was deleted.

10. Confidential memorandum, Department of State. Subject: inventory of Iranian assets in the United States, February 2, 1979, Published in the Hearings of the Subcommittee on Commerce, Consumer, and Monetary Affairs, "Federal Response to OPEC Country Investments in the United States," Part 1, p. 733.

11. Interoffice memorandum for Secretary of the Treasury Miller, March 20, 1980, published in "Federal Response to OPEC Country Investments in the United States," Part 1, p. 692.

12. Hearings before the Senate Foreign Relations Subcommittee on Multinationals, 1975, Part 5, pp. 160–61.

13. Scheuer's characterization of the Department of the Treasury's intervention is contained in a letter from Scheuer to President Carter, April 3, 1978. A review of the State Department's response to Scheuer confirms the absence of any specific or important information.

14. During this time Treasury Department officials were very worried that their efforts to suppress disclosure might not succeed. In an internal Treasury memorandum sent to Secretary Blumenthal by Assistant Secretary Bergsten on August 18, 1979, Bergsten wrote: "In the past few months we have been inundated with requests from Congress, commercial banks, the

press and private individuals for information on Saudi Arabian investments in the United States.'' Bergsten went on to note that the efforts of
Congressman Scheuer, Senator Church, and several legislators (during the
debate over the sale of F-15 fighter planes to Saudi Arabia in May 1978)
to ascertain Saudi investments had been defeated. Nevertheless, Bergsten
cautioned that because of ''concurrent pressures, there is some risk we
will not be successful.'' He urged Blumenthal to inform the Saudis of the
large number of requests for Saudi investment data that Treasury had received—and to assure them of Treasury's plans ''to the maximum extent
possible to keep the details of your assets in the U.S. confidential.'' Memo
published in ''The Operations of Federal Agencies in Monitoring, Reporting On, and Analyzing Foreign Investments in the United States,'' Part
5, pp. 199–200.

Two years before, in January 1976, Kuwaiti officials expressed their
deep concern over the ''breach of confidentiality'' when details of Kuwaiti assets held by Citibank appeared in the American press. According
to a 1979 Treasury letter I obtained under the Freedom of Information
Act, ''Secretary Simon indicated to Kuwaiti officials that the U.S. Government would do everything it could to prevent a repetition of the incident.''

15. One particularly bitter exchange between Rosenthal and Bergsten took place
on the morning of July 18, 1979, at a hearing chaired by Rosenthal:

BERGSTEN: We have provided, I think, over three hundred. The ones
we have not provided are not provided, Mr. Chairman, for three reasons. One is the reason we are discussing now—the confidentiality requirements under these acts. A second is the national security classification where, as you know, we have offered to make these documents
available if satisfactory arrangements can be worked out with the subcommittee to assure the protection of those classifications.

ROSENTHAL: Do you want me to tell you something? We do not have
to assure you of anything. We want those documents. Either give them
to us or do not give them to us—just tell us what you want to do.

BERGSTEN: I said, Mr. Chairman, that there are some documents that
we are not prepared to give to you.

ROSENTHAL: I do not blame you personally because I have a great personal respect for you but the executive branch cannot set up the rules
this government is run by. If we want the documents, we get them. If
we do not get them, we subpoena them. If you do not execute the subpoena, we hold you in contempt of Congress. The course is clearly laid
out. There is no constitutional basis for your refusing to disclose and
give to an investigating and oversight committee of the Congress the
information requested. . . . You cannot refuse in any way.

16. Unclassified summary of cable published in "Federal Response to OPEC Countries Investments in the United States," Part 1, p. 554. Though the name of the Middle East country was excised in the summary, I have learned that the country was Kuwait.
17. A glimpse into the Treasury Department's disdain for Rosenthal is provided by an internal Treasury memorandum obtained under the Freedom of Information Act. "Rosenthal is xenophobic," said the memo, "particularly with regard to OPEC investments in the U.S."
18. See the cables and documents published in "Federal Response to OPEC Countries Investments in the United States," Part 1, pp. 527–656.
19. Ibid. See especially Doc. 22, p. 528.
20. Ibid. pp. 590–91.
21. *Chicago Tribune*, May 31, 1981.
22. *Business Week*, January 23, 1978.
23. *Wall Street Journal*, September 14, 1981.
24. "Saudi Reserves under Pressure," *New York Times*, June 11, 1979, p. D1.

CHAPTER 17

1. *Time*, April 30, 1984.
2. *Washington Post*, May 15, 1982.
3. Information on the media tour is on file at the Department of Justice as part of Gray's registration as a foreign agent for the Arab Women's Council.
4. *New York Times*, January 6, 1983.
5. The owner of a private Boeing 727 and a retinue that includes a valet, barber, secretary, and host of other factotums, Talal is considered one of the world's wealthiest men: his fortune is said to exceed $6 billion. During an eighteen-month period in 1983 and 1984, Gray's expenditures on behalf of the prince included $4,000 for teddy bears as "gifts," $1,098.70 for hobby horses from Houston's Neiman-Marcus, $2,500 for chartering a whale-watching vessel off the coast of Catalina, and more than $150,000 for scores of lunches, receptions, and dinners across the country.
6. Financial data on Gray's expenditures for the League of Arab States is on file at the Department of Justice.

CHAPTER 18

1. The account of and quotes from West's talk are based upon local newspaper coverage in the *Hilton Head News*, April 21, 1983, and the *Island Packet*, April 21, 1983. In addition, I spoke with Bob Bender, a writer

for the *Hilton Head News* who reported West's address for his newspaper. In my interview with Bender, he confirmed unequivocally that West had relayed his "understanding" that Israel had participated in a massacre of "ten thousand" Christian Lebanese.

2. *Greenville News,* June 30, 1982.

3. *Kiplinger Washington Letter,* November 5, 1976.

4. Opposition to West's appointment came from the head of the Foreign Service Association, the largest organization of career foreign service officers. President Patricia Woodring declared that West was "not qualified to serve in such an important post" and that West's appointment "may be an example of political cronyism." *Columbia State,* May 25, 1977.

5. *Charlotte Observer,* January 20, 1971.

6. *Columbia State,* January 8, 1975.

7. *Charlotte Observer,* January 8, 1975.

8. *Columbia Record,* January 8, 1975.

9. *Columbia State,* August 8, 1975.

10. *Columbia Record,* April 13, 1976.

11. *Columbia State,* May 21, 1976.

12. *Columbia Record,* May 17, 1976.

13. *Columbia State,* June 26, 1977.

14. *Columbia Record,* May 25, 1977.

15. *Columbia Record,* February 23, 1978.

16. *Columbia Record,* March 29, 1978.

17. *New York Times,* March 30, 1978.

18. *Columbia Record,* April 3, 1978.

19. *Columbia Record,* April 3, 1978.

20. *News and Courier* (Charleston), March 31, 1978.

21. *Washington Post,* May 6, 1979.

22. Interview with West, February 17, 1984.

23. *Washington Post,* July 11, 1979.

24. Speech of Ambassador John West before the Forum Club in Houston, April 12, 1979. Tape of West's speech provided by the Forum Club.

25. *Columbia State,* April 9, 1980.

26. Associated Press, February 2, 1981; *New York Times,* February 3, 1981; *Los Angeles Times,* January 30, 1981.

27. *Wall Street Journal,* May 4, 1981.

28. "Tomorrow," July 20, 1981.

29. My source for the remarks West made before this group on April 12, 1984, are: a tape recording I obtained of West's talk; an account in the *Hilton Head News,* April 19, 1984; and an interview with *Hilton Head News* reporter Bob Bender.

30. *Miami Herald,* April 15, 1982.

31. Interview with West, February 17, 1984.
32. CBS, "60 Minutes," interview with Morley Safer, June 19, 1983.
33. *Miami Herald,* April 15, 1983.
34. Transcript of news conference provided by Washington reporter David Silverberg, who attended the briefing.
35. *Miami Herald,* April 15, 1982.
36. Ash Sharq Al-Awsat, February 3, 1984, translated and printed by the Foreign Broadcast Information Service, February 7, 1984.
37. This firm, according to commerce department documents, represents American companies in getting Saudi contracts.
38. A year after this conference took place, Saudi Arabia rewarded the Southern Center for International Studies. According to foreign agent registration files at the Department of Justice, Saudi agent Crawford Cook contributed $10,000 of Saudi money to the center.
39. "Proceedings of a Conference on Gulf Security," The Citadel, no. 1, Citadel papers in International Affairs; *News and Courier/Evening Post,* November 13, 1983; interviews with members of the audience; *Global Review,* Southern Center for International Studies, Spring/Summer 1982, no. 4.
40. This particular quotation is from West's speech before a conference in Birmingham sponsored by the American Arab Affairs Council, March 4, 1983.

Chapter 19

1. *Aramco World.* January–February 1977.
2. The money was never given, however, for reasons unknown.
3. *Birmingham News,* August 30, 1978.
4. *Birmingham News,* August 23, 1978; *Birmingham News,* August 16, 1978. In addition, according to a senior Birmingham businessman, Copeland and other executives arranged a meeting with Jewish leaders to calm growing fears by the Birmingham Jewish community that the prospect of trade with the Arab world was inducing an anti-Israeli disposition. The meeting, however, may have heightened Jewish fears when Copeland suggested that everyone send telegrams to President Carter "urging him not to agree to anything that is not in the best interests of the United States" right before the scheduled Camp David summit with Egyptian President Sadat and Israeli Prime Minister Begin. To some Jewish leaders, the telegram seemed like a not-so-subtle attempt to safeguard the interests of Saudi Arabia, which was worried about any Egyptian-Israeli rapprochement, insofar as the President of the United States didn't need to be told to protect "the interests of the United States."

5. *Birmingham News*, September 13, 1978.
6. AP report, *Arkansas Gazette*, December 28, 1980.
7. *World Wide Markets*, Volume 9, number 1, Spring 1982. Published by The International Department of the First National Bank of Birmingham.
8. *Birmingham Post-Herald*, July 16, 1980.
9. *UAB Report*, February 18, 1977; March 4, 1977.
10. Letter, on Birmingham Trust National Bank letterhead, from H. R. Vermilye to S. Richardson Hill, March 22, 1984.
11. Memo from Blaine Brownell, director of UAB's Center for International Programs, December 21, 1982.
12. Both Chancellor Bartlett and President Hill afterwards refused repeated telephone requests to talk to me.
13. Upon his retirement, McCloskey, in addition to his work as a lawyer, has become speaker for the Arab American Affairs Council, the American–Arab Anti-Discrimination Committee, the National Association of Arab Americans (NAAA), and other Arab lobbying groups. Since the speech he gave in Birmingham, McCloskey has delivered several variations of it around the country. In one speech arranged by the NAAA, McCloskey urged that American corporations doing business in the Middle East set aside 10 percent of all their Arab business for pro-Arab lobbying.
14. Telephone interview with H. R. Vermilye, March 22, 1983.
15. There is no suggestion that Findley engaged in or was aware of any criminal misconduct.
16. *Birmingham Post-Herald*, April 12, 1984. Arrington has also demonstrated his "evenhandedness." In 1984, he signed on as "Honorary Mission Leader" of a 1985 Birmingham tour of Israel. The tour was to kick off a Birmingham festival organized by the Birmingham Jewish community in honor of Israel.
17. The account of Young's trip to the Gulf is based upon the reporting of Paul Lieberman, "Call to Prayer Nearly Stops Mayor from Buying Wife Gift" and "Andy of Arabia," *Atlanta Journal*, March 23, 1983. Lieberman accompanied Young on the trip.
18. Unless otherwise noted, the account of the Atlanta conference is based on interviews with several participants, tape-recorded speeches, and official conference materials. The quotes of Carter's speech are taken exclusively from a tape recording of his remarks.
19. *Washington Post*, May 15, 1983.
20. *Washington Post*, May 11, 1983.

INDEX